Ultimate load design of concrete structures

Research Report

The Institution of Civil Engineers

Ultimate load design of concrete structures

REPORT OF THE RESEARCH COMMITTEE,
WITH DISCUSSION

THE INSTITUTION OF CIVIL ENGINEERS
LONDON

The Institution of Civil Engineers, Great George Street,
London SW1P 3AA

Report first published in *Proceedings of the Institution of Civil Engineers*, February 1962, pp. 339–442
Written discussion first published in *Proceedings of the Institution of Civil Engineers*, October 1962, pp. 225–269
Committee's reply first published in *Proceedings of the Institution of Civil Engineers*, November 1962, pp. 532–537
Account of informal discussion first published in *Proceedings of the Institution of Civil Engineers*, June 1963, pp. 242–248
First published together as *Ultimate load design of concrete structures* in 1964

Reprinted 1966, 1972, 1975

ISBN 0 901948 64 0

COMMITTEE

Prof. A. L. L. Baker, D.Sc.(Eng.), M.I.C.E. (Chairman), Imperial College of Science and Technology
Mr P. B. Ahm, M.Sc., A.M.I.C.E., Ove Arup & Partners
Dr S. C. C. Bate, Ph.D., B.Sc., M.I.C.E., Building Research Station
Dr W. W. L. Chan, Ph.D., B.Sc., D.I.C., A.M.I.C.E., Sir Bruce White, Wolfe Barry & Partners
Prof. R. H. Evans, C.B.E., D.Sc., Ph.D., M.I.C.E., Leeds University
Prof. M. R. Horne, M.A., Ph.D., Sc.D., M.I.C.E., Manchester University
Dr K. Hajnal-Kónyi, M.I.C.E., Consulting Engineer
Mr A. W. Hill, B.Sc.(Eng.), M.I.C.E., Cement & Concrete Association
Prof. W. T. Marshall, Ph.D., B.Sc., M.I.C.E., Glasgow University
Prof. P. B. Morice, Ph.D., B.Sc., A.M.I.C.E., Southampton University
Prof. A. D. Ross, Ph.D., B.Sc., M.I.C.E., F.R.S.E., King's College, London
Dr C. W. Yu, Ph.D., D.I.C., Ove Arup & Partners

CORRESPONDING MEMBERS

Dr Ing. R. Gartner, South Africa
Dr Eivind Hognestad, Illinois, U.S.A.
Prof. Dr. Techn K. W. Johansen, Denmark

Made and printed in Great Britain by
William Clowes & Sons, Limited, London, Beccles and Colchester

CONTENTS

Foreword 4
Introduction 5
Notation 6
Definitions 7
Design criteria for frames 9
Rigorous analysis—Method A, based on moment/curvature relation 11
Rigorous analysis—Method B, based on moment/rotation relation . 11
Simplified limit method 14
Calculation of hinge rotations 15
Calculation of permissible rotation, θ_p 16
Recommended safe limiting values of parameters influencing strength in bending and in combined compression and bending 18
Calculation of values of M and EI at L_1 18
Under-reinforced section 18
Calculation of permissible rotations at L_2 20
Alternative value of M at hinge sections 20
Alternative methods of calculating M at L_2 21
Load factor 21
Alternative recommendation of European Committee on Concrete . 24
Instability 25
Instability of columns 25
Instability of beams 27
Instability of frame as a whole 27
Recommendation on the choice of positions of hinges 28
Shear 28
Slabs 30
Shear forces 31
Application of theory to practical design of frames 33
Practical application in Denmark 33
Practical application in Russia 34
Test results 34
Co-ordination of research on inelastic deformation 34
Worked examples 36
References 1–33 44
Written discussion 47
References 34–100 89
Reply to written discussion 92
References 101–105 97
Supplementary informal discussion 98

FOREWORD

The following report has been prepared by the Committee on ultimate load design of concrete structures of the Institution of Civil Engineers Research Committee.

The Committee was appointed on 14 May, 1958, with the following terms of reference:—

1. To review existing knowledge in regard to limit design of reinforced concrete, with special reference to:—
   - (*a*) Fundamental theory,
   - (*b*) Application in design,
   - (*c*) Present codes of practice.
2. To make recommendations for future research.

It should be emphasized that the report does not attempt to act as a code of practice, although it should prove of value to the practical designer. The fundamentals of limit design are defined. Some attempt has been made to indicate a suitable form of computation curves to facilitate work in the design office. At a later stage, it will undoubtedly be possible to present simple design rules, tables, computation curves, etc., for use in designing the more common forms of structure.

In the preparation of the report, the Committee has benefited from discussion of fundamental and academic viewpoints, as opposed to the more practical considerations which are uppermost in the mind of the practising engineer.

The Committee has also benefited by close contact with the work of the European Committee for Concrete.

Recommended values of parameters for application in design have been given, and the research reports on which these are based are given in the references.

In some cases it is pointed out that further research is required and, for the time being, as a precaution, conservative values have been recommended.

ACKNOWLEDGEMENT

The Chairman wishes to acknowledge indebtedness to the Institution of Structural Engineers, the Cement and Concrete Association, and the Reinforced Concrete Association, in addition to the Institution of Civil Engineers, for their part in supporting British representation through him on the European Committee for Concrete. The meetings of the latter Committee have proved an invaluable medium of discussion and source of research information.

Introduction

ULTIMATE load theory has been widely discussed by engineers in recent years and frequently adopted as a basis of design for structural steelwork frames. The limited deformation of concrete has, however, tended to cause attention to be confined to the ultimate strength of member sections only, the distribution of bending moments being based on assumed elastic conditions. It is the purpose of this report to attempt to clear the way for a logical treatment of the structure as a whole. Few reinforced concrete structures under load behave in the elastic manner normally assumed. Beams usually crack at low loads and so their stiffness is reduced. Highly stressed sections yield but, within limits, retain their full strength. Strain measurements at working load on actual structures often bear little relation to values calculated by elastic theory, whereas ultimate loads found by tests are usually in closer agreement with calculated values.

2. To determine the true margin of safety available against failure it is essential to calculate the ultimate load. It is not, however, the intention of this report to present the ultimate load basis of design as a complete and entirely sufficient substitute for the elastic theory. The engineer should consider the behaviour of a structure and its implications at all stages of loading up to failure. The value of the ultimate load, however, must be given primary consideration, as its relation to the working load provides the only reliable measure of security.

3. However, the same general equations used to check and adjust deformations at ultimate load may also be used to obtain an elastic solution which generally need only be approximate, since it is only required to avoid load behaviour affecting serviceability, such as excessive cracking or deformation under working load. In any case, there is no point in attempting to be too precise, since the stiffness values used in the equations can only be approximate. The general elastic equations may thus be used with discretion to provide a comprehensive analysis of any frame of elasto-plastic material at both critical stages of loading. At the ultimate stage when defined by limits such as L_1 and L_2 (Fig. 1), the solution of the equations becomes a matter of simple addition.

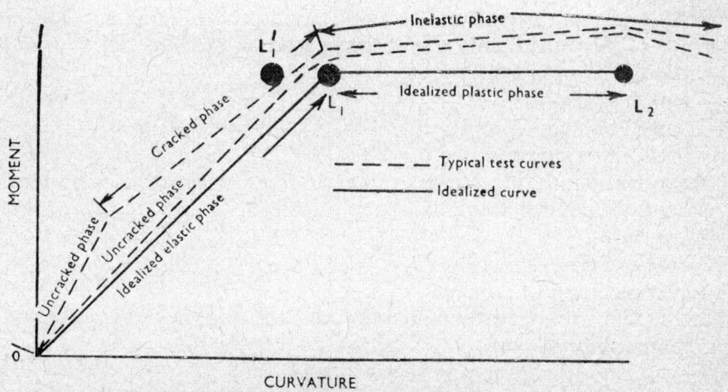

FIG. 1.—IDEALIZED MOMENT/CURVATURE RELATIONS

4. The designer, by adjusting the restraint moments at critical hinge sections or the stiffness of frame members, can readily vary, according to his requirements, the plastic deformation under ultimate load at critical sections from maximum permissible values down to zero. In the latter case, the idealized frame virtually remains elastic up to failure, but allowance must be made for the influence of cracking on the stiffness of the members. When this is ignored, as in many so-called elastic solutions, the actual bending moments may vary considerably in their distribution from that which is obtained by conventional calculations.

Notation

Unless otherwise defined in the text the following notation applies:—

L_1 = the effective yield point in concrete or steel (Figs 2 and 3) or the idealized elastic limit in a member (Fig. 1).
L_2 = the point on the idealized stress-strain curve for concrete or steel at which failure occurs (Figs 2 and 3) or the curvature limit in a member (Fig. 1).
X_k = unknown bending moment acting at hinge k.
$\bar{X}_k$ = known bending moment acting at hinge k.
δ_{ik} = rotation at hinge i due to $X_k = 1$ acting (considering bending only).
δ_{i0} = rotation at hinge i due to external load acting (considering bending only).
θ_i = resultant rotation at hinge i.
θ_p = permissible rotation at hinge i (on one side of critical section).
θ'_i = rotation at hinge i due to inelastic deformation when the initial flexural rigidity is used for elastic calculations (Fig. 5c).
$\Psi_{\bar{n}}$ = rotation at critical section $\bar{n}$ due to inelastic deformation at a section not assumed to be a hinge when the initial flexural rigidity is assumed for elastic calculations (Fig. 5c).
M = bending moment.
M_i = ordinate of bending moment diagram for $X_i = 1$.
M_k = ordinate of bending moment diagram for $X_k = 1$.
M_0 = ordinate of bending moment diagram for external loads.
E_c = Young's modulus for concrete (secant value).
E_s = Young's modulus for steel.
EI = Mr = flexural rigidity of section (based on OL_1 (Fig. 5c).
$E'I'$ = Mr = flexural rigidity of section (based on OA Fig. 5c).
r = radius of curvature.
l = length of span.
kl = length of span (Fig. 6).
b = breadth of section.
d = effective depth of section, or overall depth when there is no tension.
n_1d = depth of neutral axis at L_1.
a_1d = lever arm.
n_ud = depth of neutral axis at L_2.
c_u = cube strength of concrete.
c = concrete stress at edge of section.
C = total compression.
t_u = assumed ultimate tensile stress in steel.
T = total tension.

α_c = average concrete compression stress.
$\gamma n_1 d$ = depth to centre of compression.
e_c = strain in concrete (generally at edge of section).
e_s = strain in steel.
e_{ce} = strain in concrete at L_1.
e_{cu} = strain in concrete at L_2.
e_{se} = strain in steel at L_1.
e_{su} = strain in steel at L_2.
e_p = strain due to prestressing only (Fig. 9).
z = distance from critical section to point of contraflexure.
$\bar{l}_p$ = actual plastic length (Fig. 4c).
l_p = equivalent plastic length on one side of critical section.
k_1 = parameter for influence of type of tensile steel.
k_2 = parameter for influence of axial load.
k_3 = parameter for influence of grade of concrete.
$F = \dfrac{e_s n_1}{e_c(1-n_1)}$.
A_s = area of tensile reinforcement.
$p = \dfrac{\text{area of tensile reinforcement}}{bd}$.
$Z = \dfrac{F}{2\alpha(1-\beta)}$.
β = Value of the prestress in the steel before transfer divided by ultimate strength of steel.

Definitions

5. *Ultimate load.* The load or combination of loads which, for a particular system of loading, would cause failure of the whole structure or a part; e.g. by overturning, by the formation of a mechanism, or by rupture or weakening of the whole structure or a part.

6. In calculating an ultimate load, it is assumed that its distribution, and rate, frequency and duration of application have the worst characteristics which could occur in relation to failure.

7. *Working load.* The load or combination of loads which the structure is expected to carry, permanently or temporarily, during its useful life.

8. In determining a working load, it is assumed that its distribution, and rate, frequency and duration of application have the worst characteristics which could occur in relation to failure. (In certain instances, when supported by adequate statistical data, a load slightly less than the worst load may be accepted.)

9. *Ultimate load factor* (usually abbreviated to *load factor*). The factor by which the appropriate working load is multiplied to obtain the design ultimate load. The value may vary for different arrangements of the load system.

10. The definition in § 11 below derives from the moment/curvature relation which, in general, is as shown by the broken lines in Fig. 1. For the sake of simplicity, an idealized relation is assumed and this is represented by the two straight lines drawn in full in the figure. OL_1 is straight in Fig. 1 although curved in Fig. 2 but E_c decreases as the factor α increases.

11. *Effective yield point* (L_1 on Figs. 2 and 3). A selected point on the short-term stress/strain curve for either concrete or steel beyond which it is assumed, on

the idealized curve, that no further increase in stress can develop with an increase in strain which may occur at a plastic hinge. (See § 61 for exception.)

12. The definitions in §§ 13–15 apply to actual structures. The definitions in §§ 16–25 apply to idealized structures.

13. *Uncracked phase.* The phase in the deformation of a structural member under test before the concrete begins to crack.

14. *Cracked phase.* The phase in the deformation of a structural member under test after the concrete has cracked up to the beginning of the inelastic phase.

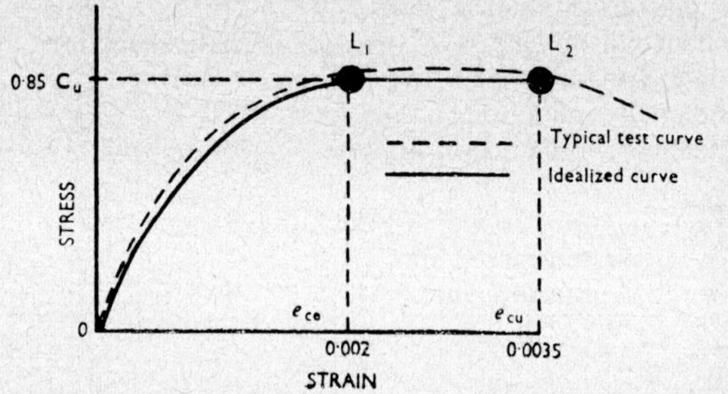

FIG. 2.—IDEALIZED STRESS/STRAIN RELATION FOR CONCRETE

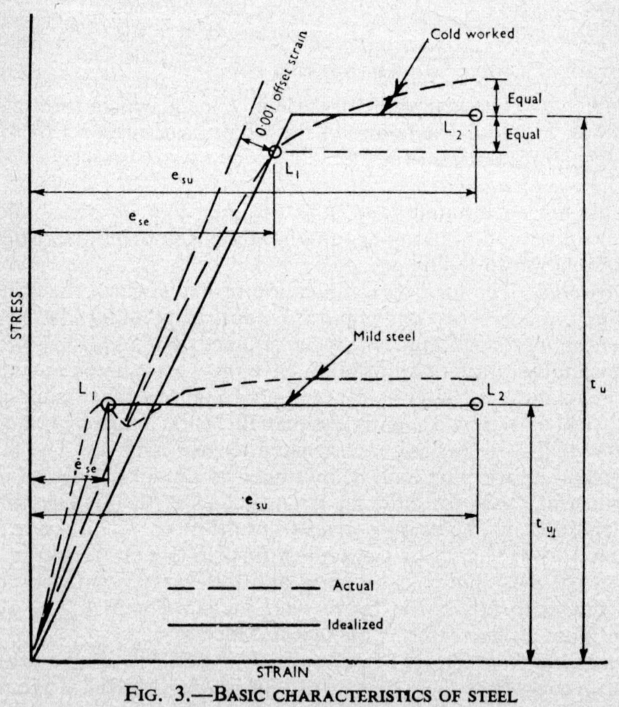

FIG. 3.—BASIC CHARACTERISTICS OF STEEL

15. *Inelastic phase.* The phase in the deformation of a structural member under test from the point at which the gradient of the moment/curvature diagram is substantially reduced to the point at which rupture occurs. Some inelastic deformation occurs in the cracked phase and a little in the uncracked phase but both are small compared with that which occurs in the inelastic phase after point L_1.

16. *Idealized elastic phase.* The phase of an idealized moment/curvature relation for a structural member subject to increasing bending moment in which there is a linear relation between bending moment and curvature. The properties of the idealized member are so chosen that its curvature is always greater than that of the corresponding actual member for any particular bending moment. (See § 61 for exception).

17. *Idealized plastic phase.* The phase of an idealized moment/curvature relation for a structural member subject to increasing deformation in which the bending moment remains constant as the curvature increases. For the same curvature, the moment carried by the idealized member is always less than that carried by the corresponding actual member.

18. *Idealized elastic limit* (point L_1 Fig. 1). The point at which the idealized elastic phase ends and the idealized plastic phase begins.

19. *Plastic moment.* The moment which obtains at the idealized elastic limit. (See § 61 for exception).

20. *Curvature limit.* The maximum curvature of an idealized moment/curvature relation for a structural member which can develop without failure. This is assumed to occur at point L_2 Fig. 1.

21. *Plastic hinge.* A section in a structural member at which the inelastic rotations are assumed to be concentrated. A plastic hinge is assumed to occur at a critical section under the plastic moment where inelastic behaviour occurs at the section and adjacent to it, so that locally a change of curvature occurs due to inelastic behaviour in the concrete or the steel or both. Its influence on the deformation of the structure is equivalent to the rotation of a hinge.

22. *Hinge rotation or angle of discontinuity.* The total change of slope due to inelastic behaviour of the concrete or the steel at and adjacent to a critical section and which, for convenience, can be assumed to be concentrated at the critical section as the rotation of a hinge.

23. *Equivalent plastic length.* The assumed length of a member over which inelastic behaviour occurs with constant curvature, so that the change of slope is equal to the actual change of slope due to inelastic behaviour (Fig. 4).

24. *Mechanism.* An articulated system of structural members connected by plastic and/or real hinges. The system is able to deform without finite increase in load until the displacements become so large that the equations of equilibrium are materially affected.

25. *Flexural rigidity* (as used in this report). The flexural rigidity at a point on the neutral axis of a member is the product of the bending moment acting about the neutral axis at the particular section and the consequent radius of curvature (see Notation EI and $E'I'$).

Design criteria for frames

26. When a frame is being designed for ultimate conditions of load, as defined, normally one application of total load applied for a short term is a

reasonable criterion and is used for the parameter values (§ 53). Special adjustments may be necessary for earthquake shock. The designer should, however, judge whether the nature of the structure and live/dead load ratio are such that some form of partial loading provides a worse condition or that a regular cycle of loading combinations may occur and cause fatigue weakening or "shakedown". The principal causes of failure to be considered are as follows:—

(*a*) The frame as a whole or in part may become a mechanism, owing to the development of a sufficient number of plastic hinges.

(*b*) A plastic hinge may fail because the curvature limit is exceeded before the structure becomes a mechanism.

(*c*) A critical section of the frame may fail in shear or bond slip.

(*d*) A member may fail in direct compression.

(*e*) The whole or a part of a structure may fail by buckling.

In some cases, combinations of these effects may lead to failure.

27. §§ 28 to 36 are not concerned with the main subject of the report which is a design method based on the idealized moment/curvature relationship shown

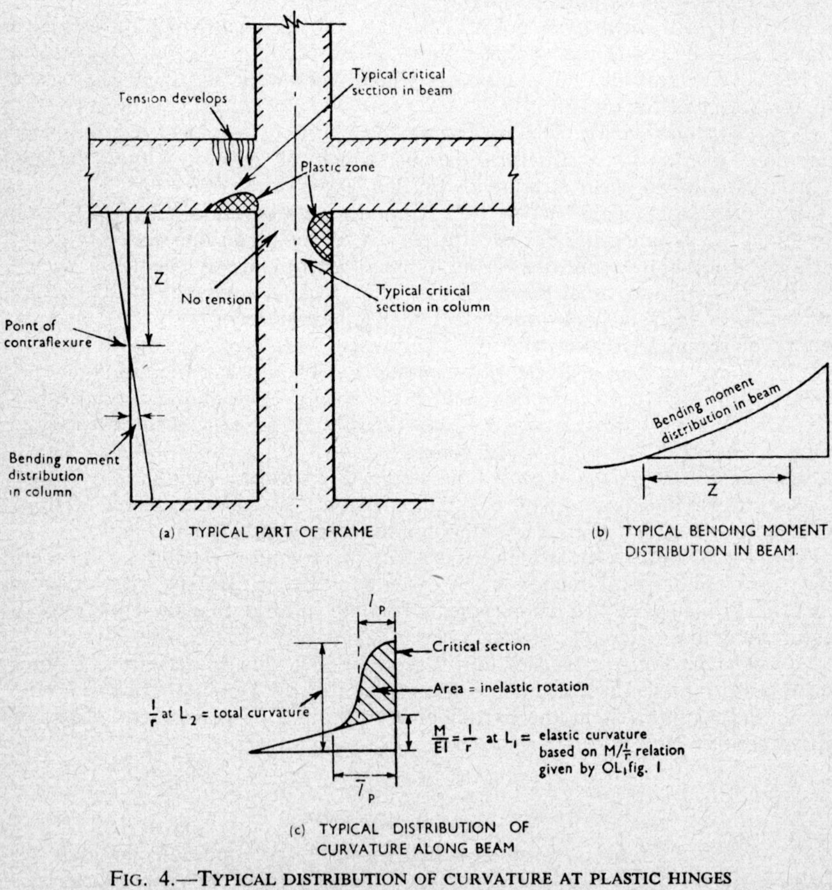

FIG. 4.—TYPICAL DISTRIBUTION OF CURVATURE AT PLASTIC HINGES

in Fig. 1, i.e. a "limit method". They have been inserted to indicate the requirements and difficulties of exact analysis. The moment-rotation method described in the second section may be of practical use in the case of continuous beam design. For the purpose of studying the "limit method", these paragraphs may be ignored. It has been assumed throughout that deformations due to axial forces, torsion and shear are negligible. An extension of the theory to include these deformations in three-dimensional frames is given in reference 33.

Rigorous analysis—Method A, based on moment/curvature relation

28. If the moment/curvature relations of all sections of a frame are precisely known, then the classical Müller-Breslau general elastic equations[1, 2] may be used. To obtain a solution of the equations, a trial and adjustment procedure may be followed until the resultant rotation values at selected hinges are zero, and the $EI\,(=Mr)$ value at each section is at the same time compatible with the resultant bending moment acting at the section. It is perhaps doubtful whether such exact solutions are practicable even for simple structures. Tests show that high local increases of curvature occur at critical sections, which in practice can only be allowed for when moment/rotation curves for at least the plastic zone are available.

Rigorous analysis—Method B, based on moment/rotation relation

29. Fig. 5a indicates part of a frame in which sufficient hinges 1 to n have been inserted for statical determinacy, hinge i at which $X_i = 1$ is assumed to act, being a typical hinge. At all hinge sections inelastic rotations $\theta'_1 \ldots \theta'_n$ and at critical sections $\bar{1}$ to $\bar{n}$ between hinges, inelastic rotations $\Psi_{\bar{1}}$ to $\Psi_{\bar{n}}$ are indicated on the tension or minimum compression side of members to facilitate the inclusion of inelastic terms in the equations. Moments M_i are also plotted in the usual way on the tension or minimum compression side of members. (Müller-Breslau equations[1, 2]).

30. When all external loads and moments X_1 to X_n act, at any section such as i the resultant rotation is zero, if all rotations due to inelastic deformations are added to those due to elastic deformations. Using the principle of virtual work to extend the Müller-Breslau classical equations, the following general compatibility equation may be established for a typical hinge section i.[28]

$$\int \frac{M_i M_0}{E'I'} ds + \sum X_k \int \frac{M_i M_k}{E'I'} ds + M_i \theta'_i + \sum M_i \Psi_{\bar{n}} = 0 \quad . \quad . \quad (1)$$

31. To obtain the distribution of bending moment at any stage of loading up to the ultimate, the M/Ψ curve (Fig. 5c) must be known for each critical section, and the X values must be found, so that all the Ψ and θ' values are compatible with the resultant moment values at the critical sections 1 to n and $\bar{1}$ to $\bar{n}$.

32. In simple cases, for example a symmetrically loaded, continuous beam (Fig. 6), it is sometimes possible to obtain without difficulty an exact solution. It is often convenient to separate the variation of the bending moments due to inelastic deformation from the distribution for elastic conditions, which is often known or easily established by standard methods.

33. Thus, if the inelastic influence only causes variations from elastic values denoted by $X_1 \ldots X_n$, values must be obtained which are compatible with

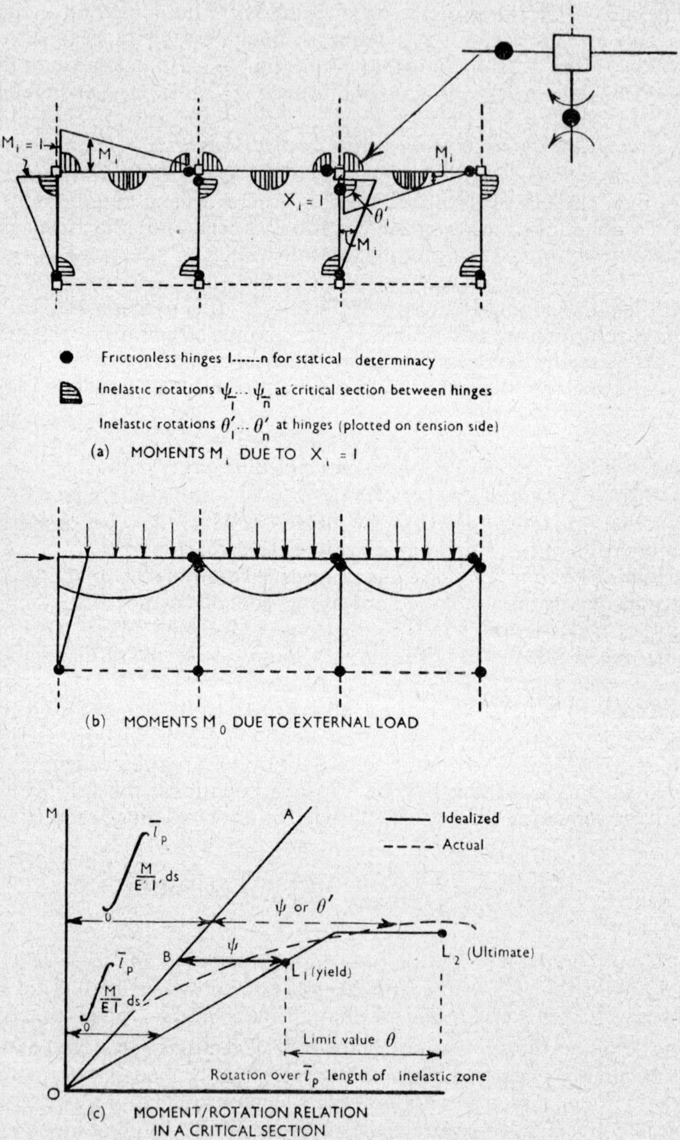

FIG. 5a, b, c.—RIGOROUS ANALYSIS DIAGRAMS

resultant bending moment values, and which satisfy the compatibility equations with the external load, i.e. M_0 terms omitted.

For the case shown in Fig. 6, omit M_0 terms:—

$$X_1 = X_2, \quad \Psi_{\bar{1}} = \Psi_{\bar{2}}, \quad \Psi_{\bar{3}} = \Psi_{\bar{4}}.$$

Equation (1) becomes

$$X_1 \int \frac{M_1 M_1}{E'I'} ds + X_1 \int \frac{M_1 M_2}{E'I'} ds - \Psi_{\bar{1}} - \Psi_{\bar{3}} = -2\theta'_1 \quad . \quad . \quad (2)$$

or

$$X_1 = 6E'I' \frac{(\Psi_{\bar{1}} + \Psi_{\bar{3}} - 2\theta'_1)}{1(2k+3)} \quad . \quad . \quad . \quad . \quad . \quad . \quad (3)$$

a result obtained by Macchi, using the theorem of three moments.[3]

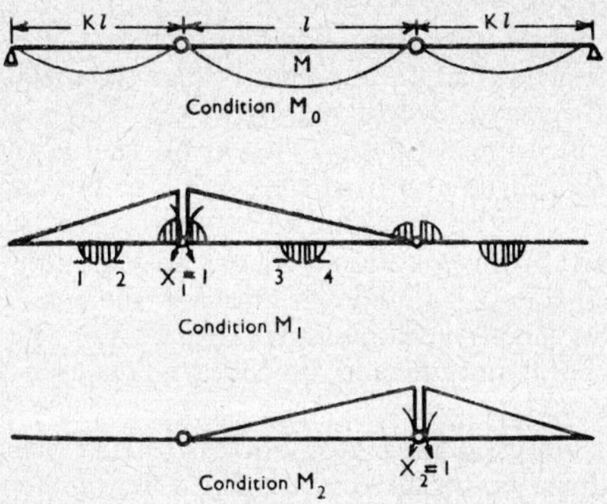

FIG. 6.—HINGES IN A CONTINUOUS BEAM

34. The values of $\Psi_{\bar{1}}$, $\Psi_{\bar{3}}$ and θ'_1 must be compatible with resultant moments at the appropriate critical sections. Such a solution is not difficult to obtain when the M/Ψ curve is known, and may be worth while in the case of structures such as bridges, in which accuracy in working stress calculations is important to avoid wide cracks to secure long-term serviceability.

35. If the moment/rotation curve (Fig. 5c) is assumed to follow the idealized line $O\ L_1\ L_2$, then in the general equation

EI must be substituted for $E'I'$
all Ψ terms become zero
θ'_i becomes θ_i

by a suitable choice of hinge positions, X terms become $\bar{X}$ terms, giving the following simplified general equation:

$$\int \frac{M_i M_0}{EI} ds + \sum \bar{X}_k \int \frac{M_i M_k}{EI} ds = -\theta_i \quad . \quad . \quad . \quad . \quad (4)$$

36. Even when the number of assumed hinges is high, by designing all members between hinges to have bending moment values less than L_1, compatibility is easily achieved by adjusting, if necessary, $\bar{X}$ values until θ values are positive and not excessive.

Simplified limit method (Based on idealized assumptions given in definitions, §§ 5–25.)

37. Since the ultimate deformation and bending strength of reinforced concrete members are subject to considerable scatter, it is reasonable, particularly for a frame which is many times statically indeterminate, to use a simplified moment/curvature or moment/rotation relation defined by limits L_1 and L_2 (Fig. 1), and to design the frame so that, under a factored load, inelastic rotations of plastic hinges lie between L_1 and L_2, and of other critical sections between 0 and L_1.

38. Frame members, except for inelastic effects at hinges, are then assumed to have constant EI value for deformations between 0 and L_1 based on the parameter values governing L_1, i.e. an assumed elastic value reduced to allow for inelastic effects.

39. Between the hinges, by design, bending moments at critical sections will be equal to or slightly less than L_1 in value. Since the inelastic deformation is small between 0 and L_1, and occurs mainly in the region of greatest bending moment, the actual inelastic rotation Ψ based on OA as the elastic moment/rotation line (Fig. 5c) will be approximately equal to but slightly less than the assumed value BL_1.

40. It can be seen from equation (4) that this assumption, while giving maximum values of θ due to possible variations in the actual relative values of Ψ and θ', could give slightly erroneous values of X. The ultimate bending strength of the member is not affected, however, unless an over-stressed section is excessively brittle.

41. The frame is designed for factored loads so that a sufficient number of plastic hinges can form to make it statically determinate without excessive deformation. Thus, when a frame is n times statically indeterminate, n plastic hinges are required and suitable positions are selected for them. The rotation of each hinge can then be calculated and, if the sign is correct, the selected position is correct. The amount of rotation must not be excessive and the members between the hinges must be designed not to yield when the nth plastic hinge forms. The moment/curvature relation of each member will then in theory be within the limits defined by the elastic phase (Fig. 1) between the hinges, and as defined by the plastic phase (Fig. 1) at the hinges. Since the members between the hinges will be designed just to remain within the elastic phase when ultimate load is applied, the load causing the formation of n plastic hinges in the structure will be almost as great as the load which produces a mechanism.

42. Where alternative systems of ultimate load may produce alternative positions of plastic hinges, each section of the frame must be designed for the greatest ultimate strength required by the various systems of load. For instance, a different set of hinges must be assumed for the alternative sway forces which may act on a frame from the left or the right.

43. Also, when a single span or alternate spans may be loaded with specially heavy live loads, such as a locomotive on a continuous beam bridge, it is sometimes difficult to obtain compatibility of deformation and bending moment values with hinges at supports. A plastic hinge should then be assumed at mid-span, and only part of the structure made statically determinate by hinges, which must be checked in regard to strength, position and degree of rotation.

In some cases with alternative spans loaded the support sections may remain elastic while the permissible value of θ develops at the mid-span hinge.

CALCULATION OF HINGE ROTATIONS

44. Since members of the frame, except at hinge sections, are designed to remain within the idealized elastic phase up to ultimate load, as defined, standard general elastic equations[1, 2] may be used to calculate deformations. The usual unknown terms X become known terms $\bar{X}$, and the value of θ at each hinge may be found by summing the left-hand terms of each equation. All these have known values.

45. Design criteria ensure that, before failure occurs, n plastic hinges will form in a frame which is n times statically indeterminate. The rotation θ_i of hinge i is obtained from the typical general equation:

$$\delta_{i0}+\sum\bar{X}_k\,\delta_{ik} = -\theta_i \quad \text{(Ref. 4, Fig. 7)} \quad . \quad . \quad . \quad . \quad (5)$$

where

$$\delta_{ik} = \int\frac{M_iM_k}{EI}ds \quad \text{(Ref. 1, 2)} \quad . \quad . \quad . \quad (6)$$

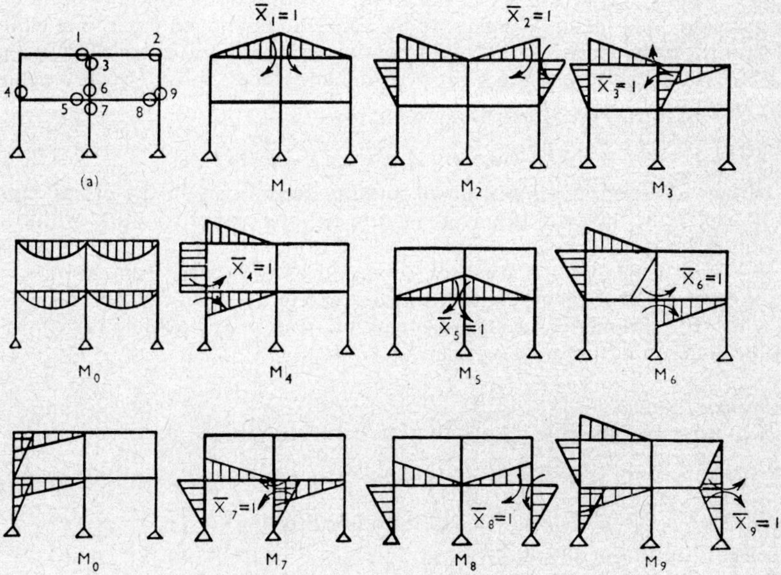

FIG. 7.—HINGE POSITIONS AND MOMENTS IN A FRAME

46. It is assumed as already stated that members between hinges have a linear moment/curvature relation. The value of EI must therefore be assumed to be constant, and is conveniently assumed equal to the value at the "idealized elastic limit". In calculations, a safe minimum value of EI is assumed in order to obtain near maximum deformations. In uncracked members, such as

columns, uncracked values of EI are assumed. In cracked members, such as beams, the value of EI at the cracked critical section is assumed.*

$$\text{Change of slope} = \int \frac{e_c}{n_1 d} ds \qquad \text{which} = \int \frac{M}{EI} ds$$

where $E = \dfrac{c}{e_c}$ = secant modulus for concrete

$$I = \frac{M n_1 d}{c}$$

where values of M and n_1 apply at L_1 of the moment/curvature diagram (Fig. 1) and of e_c and c at L_1 (Fig. 2).

In a cracked rectangular section

$$EI = \frac{\alpha c}{c_u} \frac{1}{e_c} (n_1{}^2 - \gamma n_1{}^3) b d^3 c_u \quad . \quad . \quad . \quad . \quad . \quad . \quad (7)$$

47. For safe short-term values of parameters see § 53. Under long-term loading, deformations increase, but consideration of short-term loading can provide a satisfactory criterion for safety, since it has been shown[5] that isolated members designed on a basis of parabolic distribution of concrete stress have an adequate load factor of safety where over-load is applied for a long period. Moreover, under such a condition, non-linear creep will often cause advantageous redistribution of stress in a redundant frame, except where buckling may occur.

Calculation of permissible rotation, θ_p

48. In idealized structures, hinge rotation occurs only in the plastic phase, and hence θ_p is the total ultimate rotation less the elastic rotation, within the plastic hinge zone.

49. Fig. 4c shows the form of curvature as measured from tests, which develops at a plastic hinge, and an idealized representation of the same behaviour in which for simplicity, an equivalent plastic length l_p is assumed over which plastic rotation occurs with constant curvature.

Empirical formula

50. Case 1 (Figs 8a and 8b): Tension occurs at critical section

$$\theta_p = \frac{(e_{cu} - e_{ce})}{n_u d} l_p \quad . \quad . \quad . \quad . \quad . \quad . \quad . \quad . \quad (8)$$

(on one side of critical section)

Case 2 (Figs 8c and 8d): No tension

$$\theta_p = \frac{(e_{cu} - e_{ce})}{d} l_p \quad . \quad . \quad . \quad . \quad . \quad . \quad . \quad . \quad (9)$$

(on one side of critical section)

---

* Occasionally for precise calculations, e.g. in connexion with research, it may be necessary to assume EI values based on a lower limit L'_1 (Fig. 1) for parts of the frame for which increased stiffness increases θ values, and conversely to investigate possible alternative positions of hinges.

In both cases, limiting values recommended as suitable for design are as follows:

$e_{ce} = 0{\cdot}002$

$e_{cu} = 0{\cdot}0035$ unbound concrete

$e_{cu} = 0{\cdot}012$ well bound concrete[4, 31]. When used a careful check should be made on serviceability requirements, § 72.

$$l_p = k_1 k_2 k_3 \left(\frac{z}{d}\right)^{\frac{1}{4}} d$$

$$k_2 = \left(1 + 0{\cdot}5\frac{P}{P_u}\right)$$

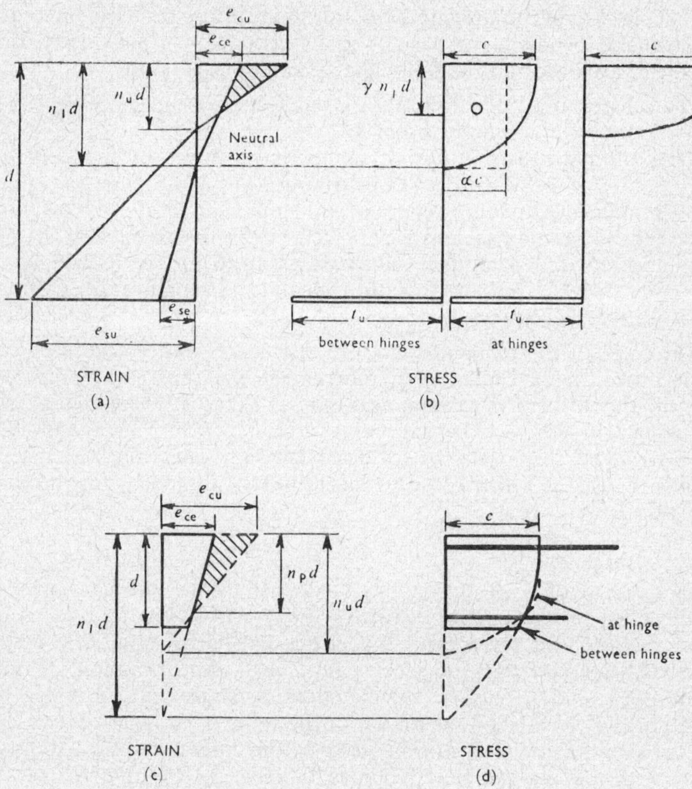

FIG. 8.—TYPICAL STRESS AND STRAIN DISTRIBUTIONS

P_u = ultimate capacity of the member for axial load when no bending moment acts.

P = ultimate axial load for the member (allowing for the bending moment when present).

51. Test results[32] available so far show that the following k values give

reasonable and safe compatibility between measured values of θ and values calculated by equations (8) and (9):

$k_1 = 0{\cdot}7$ mild steel $\qquad k_1 = 0{\cdot}9$ cold worked steel
$k_3 = 0{\cdot}6$ when $c_u = 6{,}000$ lb/sq. in. $\qquad k_3 = 0{\cdot}9$ when $c_u = 2{,}000$ lb/sq. in.
(intermediate values by proportion).

52. The limits of the values of $\frac{z}{d}$ and $\frac{l}{d}$ which occur in practice in a broad range of structural members limit l_p to a range of about $0{\cdot}4d$ to $2{\cdot}4d$.

Recommended safe limiting values of parameters influencing strength in bending and in combined compression and bending

53. Calculations of the strength of sections in resisting bending, or combined compression and bending, for various grades of concrete and kinds of steel reinforcement, are based on the following assumptions (Fig. 8).

(*a*) The distribution of strain across the section is linear (except for in fluence of F when not equal to 1).
(*b*) The ultimate distribution of compression stress in the concrete is parabolic (or an equivalent distribution).
(*c*) The moment/curvature relation for members is ideal'zed, as shown in Fig. 1. Limits L_1 and L_2 are defined by the characteristics of the concrete and steel, and values are assumed for the influential parameters which ensure that the calculated strength and deformation are safe as compared with test results.

54. The form of the moment/curvature test curves varies according to the type of steel used, the percentage of reinforcement, the transverse binding of the concrete, the degree of prestressing, and the axial load. These characteristics may be studied in Ref. 7. Typical curves are shown in Figs 1 and 10 and compared with the corresponding idealized curves. The values of the parameters given in Tables 1 and 2 lead to safe values of L_1 and L_2 for the various cases mentioned.

Under-reinforced section

55. An under-reinforced section may be defined as one in which L_1 for steel (Fig. 3) develops under bending before L_1 for concrete (Fig. 2). In an over-reinforced section, the reverse occurs. In a balanced design, L_1 develops simultaneously for concrete and steel. As a safe approximation, M for an under-reinforced section may be calculated as $A_s t_u a_1 d$, with the value of a_1 assumed to be that for a balanced section. The value of n_1 assumed to calculate EI from the expression $EI = E_c M n_1 d/c$ must be the correct value for the under-reinforced section when the steel strain is L_1 (Fig. 3). Within the idealized elastic range (Fig. 1), the moment/curvature relation has been assumed to be linear, although the idealized stress/strain curve for concrete is as shown in Fig. 2. This is justified, because $E_c\alpha/F$ is almost constant, which is evident from bending tests on beams over-reinforced with mild steel. In such tests, while the steel strains elastically, the neutral axis depth remains constant up to a strain of 0·002 in the concrete, which indicates that, as α increases from 0·5 to about 0·7, E_c/F decreases at the same rate. For a given value of n, therefore, the value of EI may be assumed to be constant within the idealized elastic range.

TABLE 1.—CALCULATION OF VALUES OF M AND EI AT L_1 (FIG. 1)
CASE 1: TENSION OCCURS (FIG. 8a, 8b)

Application	Parameter	Safe limiting value	Remarks
n_1	e_{cc}	0·002	
	e_{se}	L_1	L_1 from test results (Fig. 3)
C and a_1	$\alpha\frac{c}{c_u}$	0·60	$c_u < 3{,}000$ lb/sq. in.
		0·55	c_u 3,000–5,000 lb/sq. in.
		0·50	$c_u > 5{,}000$ lb/sq. in.
	α	0·67	when $e_{cc} = 0{\cdot}002$
	γ	0·40	See Fig. 8b
			Alternatively, a parabolic distribution of stress applicable to any shape of section, as in Fig. 8b, or an equivalent rectangular stress block, as in C.P. 114*
	c	$0{\cdot}85c_u$	To be used in conjunction with a parabolic stress distribution
T	t_u	L_1	See Fig. 3 from test results
	Fig. 8b	L_2	See Fig. 3 cold-worked steel, HINGE SECTIONS ONLY ref. §62
$I = \frac{Mn_1d}{c}$	n_1		Value at which the safe limiting value of either e_{cc} or e_{se} occurs
	c		,,
	M		,,
$E_c = \frac{c}{e_c}$	c	$0{\cdot}85c_u$	Gives value of E_c at L_1 (Fig. 2)
	e_c	0·002	
n_1 or n_u	F	1·0	Good bond. Small diameter plain bars
		0·8	Poor bond. Large diameter plain bars

* British Standard Code of Practice 114:1957

It is convenient, therefore, to calculate EI from the concrete stress distribution at a strain of 0·002. The correct value of n_1 must be used for the percentage of steel and amount of pre-strain present. The value of E_cM/c however, may be based on safe limiting values of stress and strain for concrete in compression

TABLE 2.—CASE 2: NO TENSION OCCURS (FIGS 8c, 8d)

Application	Parameter	Safe limiting value	Remarks
n_1	e_{ce}	0·002	
	c	$0{\cdot}85c_u$	To be used in conjunction with a parabolic stress distribution
I			Equivalent concrete section
E_c	c	$0{\cdot}85c_u$	Gives value of E_c at L_1 Fig. 2
	e_c	0·002	

at L_1, but using the correct value of n_1 for the section. For example, in the case of an under-reinforced rectangular section, assume:

$$E_c = \frac{c}{e_c}$$

$$M = \alpha c\, bd^2(n_1 - \gamma n_1^2)$$

so that

$$EI = \frac{E_c M n_1 d}{c} = \frac{\alpha c b d^3}{e_c}(n_1^2 - \gamma n_1^3)$$

Calculation of permissible rotations at L_2 (Fig. 1)

56. Except for special cases (§ 61), it is assumed that the value of the bending moment or combined moment and axial force at L_2 is the same as at L_1, and calculated values are based on parameter values for L_1.

57. This is not strictly compatible with the change in value of n_1 from L_1 to L_2 and an assumed constant value of c and shape of stress distribution. However, the bending moment or combined moment and axial force does not greatly increase in value between the idealized points L_1 and L_2, and the assumption is on the safe side, see Table 3.

58. The strain limits (Table 3) are recommended as suitable for ensuring adequate stiffness of frame members to avoid excessive hinge rotations. Lower values of e_{ce} may be used for under-reinforced beams and calculated from the stress in the concrete at L_1 (Fig. 2).

59. When the tensile steel is prestressed, the idealized stress/strain curve must be transformed to an assumed origin O', a point on the curve at which the strain is equal to e_p, the strain due to prestressing only (Fig. 9).

60. Stress values in compression steel are based on strain values, deduced from a linear distribution of strain across the section, in conjunction with the stress/strain curves for the steel.

Alternative value of M at hinge sections

61. When cold-worked steel is used, or a plastic hinge section is over-reinforced or remains in compression throughout, the actual moment/curvature diagram may rise between L_1 and L_2, so that to assume the value of M at L_2 to be greater than L_1 is justified. The idealized moment/curvature diagram may

then be assumed as in Fig. 10. Between hinges, the bending moment must not exceed the value at L_1 as already defined. Bending moment values at L_2 may be based on the following:

Alternative methods of calculating M at L_2

62. Either the parameter values of Table 5 may be used or the value of M at L_2 may be taken to be the mean of the value at L_1 and the ultimate value given by a moment/rotation curve obtained by test. The mean value is assumed as a safeguard against brittle sections at other associated critical sections failing before the full ultimate deformation can develop at the hinge section.

Load factor

63. A stress factor of safety of 2 in regard to reinforcement is normally satisfactory for reinforced concrete design, and is accepted as a basis for the recommendation of B.S. CP 114. In B.S. CP 115, for prestressed concrete design, load factors of 1·5 on dead load and 2·5 live load are specified, with an

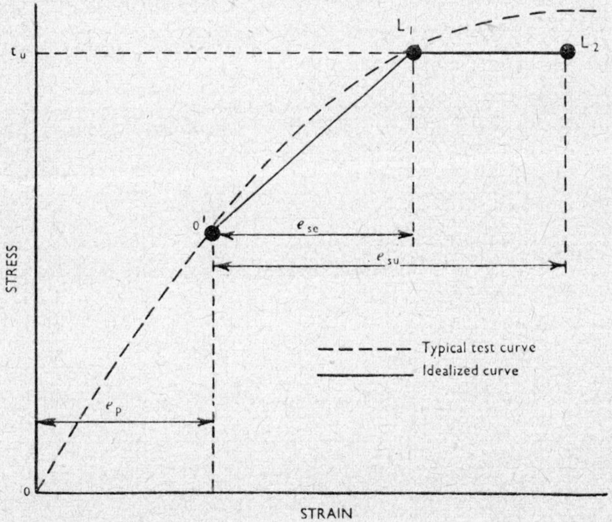

Fig. 9.—Typical and idealized stress/strain relation (prestressed steel)

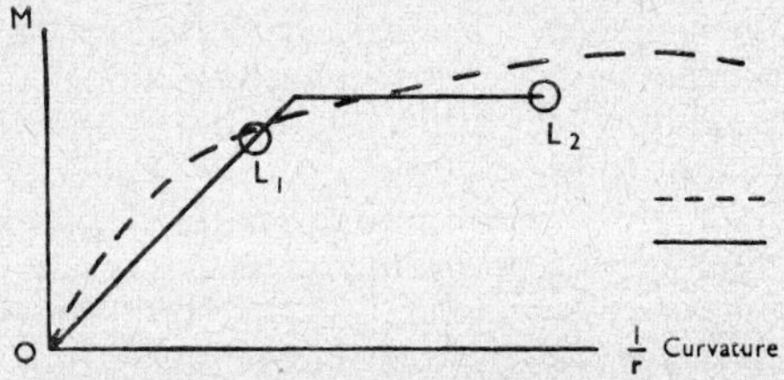

Fig. 10.—Typical and idealized moment/curvature relation

TABLE 3.—CASE 1: TENSION OCCURS (FIGS 8a and 8b)

Application	Parameter	Safe limiting value	Remarks
n_u from $C=T$	c	$0{\cdot}85c_u$	Stress distribution as at L_1
	t_u	L_1	See Fig. 3 from test results (exception see §61)
θ_p values	e_{cu}	0·0035	Unbound concrete
	e_{cu}	0·012*	Well bound concrete (details Ref. 4)
	e_{su}	L_2	L_2 from test results (Fig. 3)
l_p values			See § 50

TABLE 4.—CASE 2: NO TENSION OCCURS (FIGS 8c AND 8d)

Application	Parameter	Safe limiting value	Remarks
θ_p values	e_{cu}	0·0035	Unbound concrete
	e_{cu}	0·012*	Well-bound concrete
l_p values			See § 50

* Strains are within the binding and only the bound core of the concrete may be assumed in the calculation.[31]

TABLE 5

Application	Parameter	Safe limiting value	Remarks
n_u from $C=T$	e_{cu}	0·0035	
	e_{su}	L_2	L_2 from test results (Fig. 3)
C and a_1	c	$0{\cdot}85c_u$	Parabolic distribution of stress

overall limit of 2 on combined load. There are, however, further safeguards in both codes in regard to the design of over-reinforced sections.

64. When certain conditions are particularly favourable, it is reasonable to use a smaller value than 2, provided conditions (*a*) to (*d*) § 68 are applied to the calculations. The formula given below is a device for determining a reduced value down to a lower limit of 1·25, according to various conditions which may exist, some of which influence the probability of failure. These influences are weighted with an allowance for the seriousness of failure.

65. A small minority of the committee, including Professor M. R. Horne,

opposed the principle of permitting reductions in the factor of safety value on grounds of expediency.

66. *The load factor* (L.F.) should be determined by assessing the relative importance or weight (W) of the following six parameters and then applying the formula:—

$$\text{L.F.} = \frac{\sum W}{32} + 1.25$$

The parameters to be considered are as follows:

(*a*) *Seriousness of results of failure* (*human or economic*)
Assess weight between 0 (not serious) and 8 (very serious).

(*b*) *Workmanship*
Assess weight between 0 (very good) and 4 (occasionally bad).

(*c*) *Load conditions, including consideration given to combined loads.*
Assess weight between 0 (accurate and controlled) and 4 (less accurate and uncontrolled).

(*d*) *Importance of the member in the structure*
Assess weight between 0 (unimportant) and 4 (vital).

(*e*) *Warning of failure*
Assess weight between 0 (ample) and 3 (none).

(*f*) *Loss of strength, including provision for maintenance*
Assess weight between 0 (none likely) and 1 (maintenance normal).

67. In using this method, calculations are to be made by the ultimate load theory which, when the strength of the materials, intensity of loading, and conditions of ultimate support and eccentricity are correctly assumed, is likely to ensure that assessments of ultimate strength have a maximum error of 15%. In such cases, the value of L.F. for concrete structures will lie between 2 and about 1·25.

68. The following conditions also must apply to the calculations:—

(*a*) The calculations should be based on safe limiting values of the strength of the concrete and steel, which can be assumed to be $m-2{\cdot}33\sigma$

where m = mean strength

and σ = the standard deviation of a large number of test results and is given by

$$\sigma = \sqrt{\frac{\sum(x-\bar{x})^2}{n-1}}$$

where n = the number of results

x = the individual result

$\bar{x}$ = the arithmetic mean of the results.

(*b*) The loads should be maximum values based on statistical surveys or some reliable form of load control.

(*c*) The calculations should allow fully for ultimate value of eccentricity of loading, movement of the supports or other factors which might influence the behaviour of the structure.

(*d*) The calculations for ultimate strength should include appropriate allowances for strengthening when these are necessary to allow for fatigue, and increased cross-sectional areas to compensate for exceptional wear or corrosion that may occur during the life of the structure.

69. The formula gives a value of L.F. which is adequate when all factors, which may be given definite values or statistically based values, are properly used in the calculations. It is important that safe limiting values of the strength of the materials should be chosen, and not mean values, and that the maximum values of loading should be adopted. In this way allowance, based on statistical data, is made for the standard of quality control to be achieved and the degree of load limitation which may be expected.

70. The minimum strength ($m-2{\cdot}33\sigma$) assumed allows about one test specimen in a hundred to be below the required minimum, while the maximum load assumed may not prevent about one structural member in a hundred from being slightly overloaded. A load factor of 1 would therefore probably result in the failure of about one structural member in a number between 100 and 4,000, if the weak material occurs at the critical part of a critical section. Normally a load factor of 2 is required in Codes of Practice to provide an additional margin. Experience has shown that this is adequate when minimum strength and maximum load are included in the calculations. Moreover, a value of 2 requires an acceptable standard with sufficient margin to ensure that sub-standard material is fairly obvious to an inspector at the time of construction and is rejected, or that any load exceeding the permissible maximum is obvious to a user. Such a margin, therefore, excludes all possible causes of failure except the unexpected and extraordinary which cannot be allowed for. It also should allow a 25% overload test to be carried out, if desired without permanent damage. Where conditions are particularly favourable, the use of the formula with the appropriate weightings will provide a factor which is less than 2.

Alternative recommendation of European Committee on Concrete

71. *Values for the load factor.* The basis of assessment for factors of safety as agreed at Rome in 1957, is as follows:

Notation

Materials

R_d = safe limiting strength assumed in calculations

R_c = characteristic strength assumed to be the arithmetic mean of the lower half of all the results from at least 20 tests

c_m = factor depending on the probability of the concrete or steel in the structure being weaker than R_c

Loads

s_d = safe limiting load assumed in calculations

s_m = mean of maximum values of loads from statistical investigation of similar structures

s_c = characteristic load

K = factor depending on the accepted probability of loads being greater than s_c (sometimes less than s_c, e.g. for stability)

σ = unitary standard deviation of observed loads

c_s = factor depending on the maximum acceptable probability of failure.

Materials

$R_d = \dfrac{R_c}{c_m}$ $c_m = 1{\cdot}15$ to $1{\cdot}2$ for steel; $c_m = 1{\cdot}5$ to $1{\cdot}6$ for concrete; $R_c =$ cylinder strength — for accurate calculations and good control on site.

Loads

$s_c = s_m(1 \pm K\sigma)$ $K = 1$

$s_d = s_c\, c_s$ $c_s = 1{\cdot}25$ for accurate calculations and good control on site.

Combined loads, e.g. wind and snow loads, together with effects due to temperature or shrinkage, may be reduced by 20%, provided the result is not less than any single load.

Values for the primary load factor

$= c_m \times c_s = 1{\cdot}5$ for steel
$2{\cdot}0$ for concrete

Adjustments

Resulting damage very serious		+15%
Resulting damage very slight		−15%
Calculations of average accuracy	c_s	+15%
Average quality control	c_m	+15%

Design for serviceability

72. In structures in which it is specially important to avoid cracking under working load a solution of the compatibility equations (5) should be obtained in which values of θ are approximately equal to 0. Working stresses based on the distribution of bending moments obtained can then be calculated, and a check made to ensure that excessively wide cracks will not occur. A similar check may be necessary sometimes when excessive deformation or stresses which cause fatigue weakening may occur at working load. For continuous beam bridges it may be worth while being more precise in regard to stiffness values at working load, by using the Rigorous Method B, § 29.

INSTABILITY

73. The problem of the instability of frames may be sub-divided as follows:—

instability of individual columns,
instability of individual beams, and
instability of the frame as a whole due to sway.

Instability of columns

74. Columns in frames are usually subjected to bending as well as to direct load. The bending moment in such columns obtained from preliminary consideration of the hinge positions will need to be increased to allow for the deflexion of the column under load.

75. When sway forces are negligible, a column may be designed in the normal way as a slender column, including allowance for the worst arrangement of vertical live load. When sway forces are significant, let the full line in Fig. 11a

be the bending moment on the column carrying a vertical load P, assuming plastic hinges at top and bottom and neglecting deformations both of the column and of the structure. The deflexion of the column alone causes an additional bending moment of $P \times$ deflexion, and this additional bending itself causes a further simultaneous deflexion and an additional moment, the total moment due to these two causes being of the form shown in Fig. 11b.

76. The side-sway of the building also causes an additional moment of $P \times$ the deflexion due to this cause. This additional moment in turn produces further simultaneous deflexion and a further moment, the total moment due to these two causes being of the form shown in Fig. 11c.

77. These moments cause a movement of the position of the plastic hinges from the ends inwards and the rotation of the hinges in these new positions add further moments of the form in Fig. 11d to the moment in the columns.

78. The process of determining the deflexion bending moments and ultimately

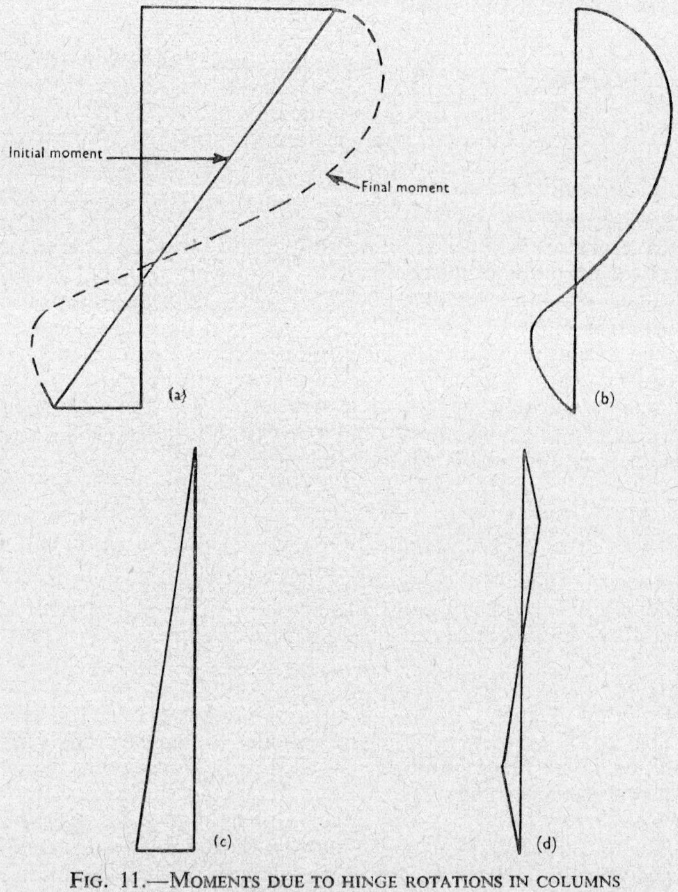

FIG. 11.—MOMENTS DUE TO HINGE ROTATIONS IN COLUMNS

the hinge rotations is largely one of estimation, trial and error. The final solution must, however, include (1) direct load on the column, (2) bending moment on the column, (3) rotation of the hinges, and the combined effects of these must be within the safe limits.

79. In addition to the above, it may be necessary to allow for deflexions due to errors of construction or lack of uniformity in the stiffness of the concrete across the column section by increasing the eccentricity of the direct load discussed above.

80. In the trial and adjustment method of checking the actual positions and rotations of the plastic hinges in columns, taking into account all the effects mentioned above, the principal steps are:

(*a*) assume magnitude of deflexions and the consequent moment diagram of the column of the kind shown in Fig. 11;

(*b*) assume resultant shifting of plastic hinges in a column as shown in Fig. 12, for a typical rectangular frame, which gives the moment diagram due to unit moment applied at the hinges.

The hinge rotations of the frame can be worked out as in the normal case.

81. The agreement of the resultant deflexion with the values assumed in (*a*) and (*b*) can be checked. If this is not satisfactory, another set of values of (*a*) and (*b*) may be used and the process repeated as demonstrated in Reference 4. The design is safe as long as the calculated resultant deflexions are less than the assumed values.

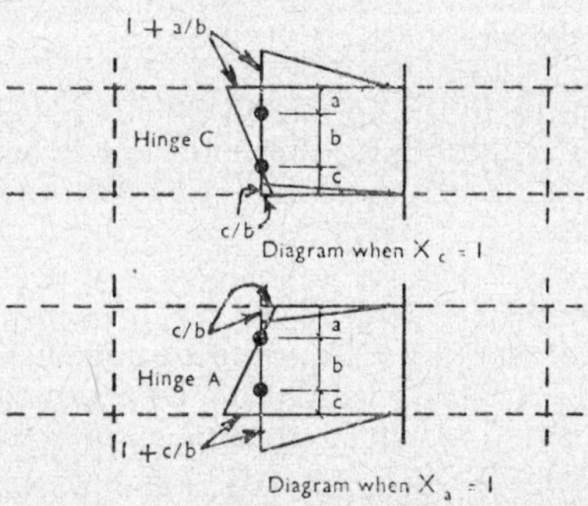

FIG. 12.—MOMENT DIAGRAMS DUE TO UNIT MOMENT APPLIED AT THE HINGES

Instability of beams

82. The assessment of the instability of reinforced concrete beams is a very complex analytical problem, and few experimental results are at present available. It cannot be assumed that any formulae derived for homogeneous sections will apply to reinforced concrete beams. As they possess comparatively high torsional rigidity, there is less danger of buckling in reinforced concrete beams than in structural steel sections. It seems that, unless a beam is exceptionally highly reinforced, lateral instability seldom occurs before flexural failure, as suggested by some preliminary tests conducted by Hansell and Winter.[8]

Instability of frame as a whole

83. Under most cases of loading, a reinforced concrete frame may be

reinforced so that at least one corner per floor remains continuously elastic and places a known limit on the sway even after the *n*th hinge has formed (Fig. 7). A typical case is shown in Fig. 13. It is possible in reinforced concrete structures to avoid the type of instability observed by Wood[9] in steel structures.

Recommendation on the choice of positions of hinges

84. When a set of hinges is to be chosen for the calculation of ultimate load and hinge rotations, as emphasized in § 44 onwards it is best to assume these n hinges to be situated at the junctions of members. In an ordinary frame, the hinges would be assumed at places as shown in Fig. 7.

85. If this is done, the influence coefficients in equation (5) can be readily

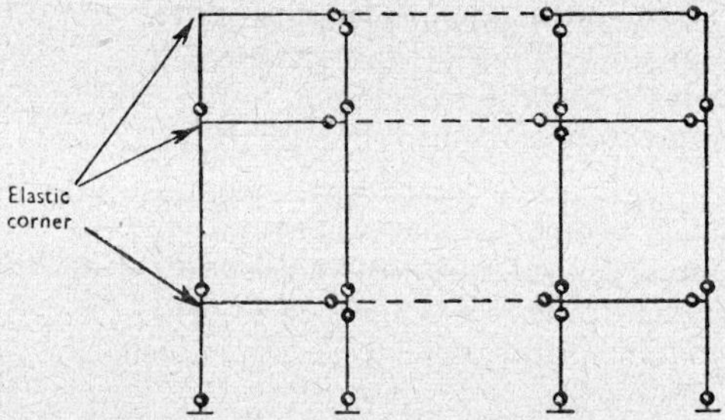

Fig. 13.—Corners remaining elastic even after formation of the nth hinge

computed. Further, owing to the normal shape of bending moment diagrams under ordinary loading conditions, the recommended choice of hinges would give smaller deflexion of the structure as a whole at working load than the alternative arrangement.

86. For framed structures, unlike continuous beams, it is often difficult to recommend certain sectional criteria, such as l/d and l/b ratios to ensure tolerable deformations at working load. In such circumstances, if it is desired to check approximately the deformations to be expected, the working load moment distribution can be obtained by simple relaxation of equation (1) to give $\theta=0$.

87. When this is done, certain block relaxation procedure may be advantageously employed, especially when the structure is highly redundant.

88. If the hinges in Fig. 14 are grouped into three such types such as "A", "B", and "C", then the recommended relaxation procedure would be as follows:

(*a*) Block relax hinge moments of hinge type "B"
(*b*) Block relax hinge moments of hinge type "A"
(*c*) Block relax hinge moments of hinge type "C"

Procedures (*a*), (*b*) and (*c*) should begin with the top storey and be repeated for the floor below until ground floor is reached.[4, 27]

Shear

89. The development of flexural failure with its consequent deformation forms the main basis for the ultimate load method of design for concrete frames.

Modes of failure which are brittle in character and interfere with the rotational capacity of hinges must be avoided. In general, failures due to shear are accompanied by little deformation and it is therefore essential that adequate shear resistance be provided in design. At the present time, the design procedures for resisting shear, although based on tests to failure, are not entirely satisfactory, and further research is necessary before these can progress beyond the recommendations of the existing codes of practice for reinforced and prestressed concrete (CP 114:1957 and CP 115:1959). It is not possible to deal with detailed design for shear in this report. Nevertheless, it is considered desirable to set out some of the factors that should be taken into account when assessing shear resistance.

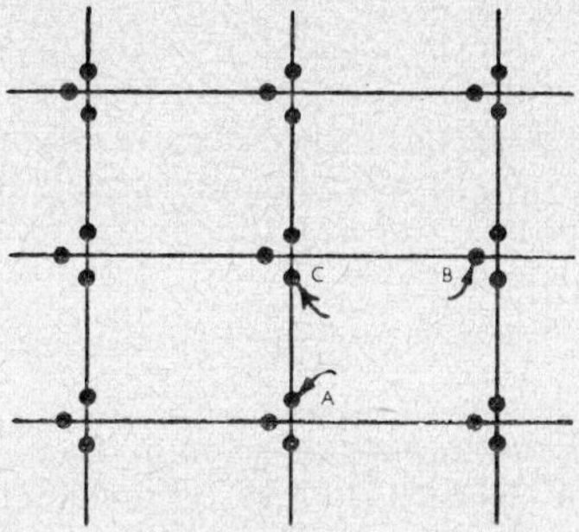

FIG. 14.—POSITIONS ASSUMED FOR HINGES IN AN ORDINARY FRAME

90. It is recommended in the code of practice for prestressed concrete (CP 115:1959) that all major structural members should have shear reinforcement, irrespective of the allowances for permissible stresses in the concrete which might suggest otherwise. This recommendation should be applied to all types of concrete frames.

91. The calculation of the ultimate moment of resistance of concrete sections in flexure is based on assumed minimum values for the strengths of the concrete and of the steel. The strength and stiffness of these materials in the actual construction may be greater, and thus the actual ultimate moment may be greater than the calculated moment. A framed structure may therefore be capable from flexural considerations of supporting a greater load than the calculated ultimate load. In order that adequate warning of collapse may be provided, the shear resistance of all parts of the structure should be sufficient to ensure that, if failure occurs, it is by bending and not by shear. For cold-worked reinforcement, which has no clearly defined yield-point, the proof stress adopted in design is usually less than the actual stress developed in steel at failure of the beam. This may lead to underestimation of ultimate load. It should also be borne in mind that the formation of plastic hinges may reduce the shear resistance of a reinforced concrete member. Further research on this subject should be carried out.

92. Shear failure may be initiated by failure of the anchorage of reinforcement. In particular, the length of bar continued beyond the point of contraflexure should be sufficient for effective anchorage, consideration being taken of both the elastic distribution of moments and the re-distribution of moments at

ultimate load. If deformed bars are used, it is essential that adequate transverse reinforcement should be provided in beams and frames at laps and ends of bars to avoid bursting of the concrete by the deformed bars.

93. In statically indeterminate concrete frames, axial loads may occur in beams. These loads may be tensile in character as the result of shrinkage under working conditions, or they may be tensile or compressive in character and due to severe deformation of other parts of the structure under ultimate load conditions. Such loading may reduce the shear resistance of the beams.

94. Statically indeterminate structures generally have a greater reserve of strength against bending failure than statically determinate structures as a result of the re-distribution of bending moments. This greater reserve does not, however, exist with regard to failure in shear, since the calculated value may be less than the actual value because of excessive local stiffness in bending. While the provision of shear reinforcement is always desirable even when shear stresses are low, it is necessary when shear stresses are high and shear reinforcement is provided to limit the stresses in the concrete to avoid diagonal compression failures; in both these respects, the present recommendations of codes of practice are not adequate.

Slabs

General

95. This section concerns failure of slabs in bending. The ultimate load design of reinforced concrete slabs is based on the applied yield-line theory for slabs. The applied yield-line theory is developed from the pure rupture line theory which is a special case of the general theory for rupture.

96. The condition for rupture for slabs is that the bending moment has reached a maximum value m or minimum value $-m'$ at a sufficient number of points to cause the formation of a mechanism. This is the case when the maximum or minimum values are reached at a definite pattern of yield lines extending over the surface of the slab.

97. The applied yield-line theory is based on the following assumptions:

(*a*) that the slab follows the same idealized moment/curvature relation as other concrete structural members, i.e. with a distinct elastic phase and a distinct plastic phase;

(*b*) that the elastic deformations are negligible in comparison with the plastic ones.

98. The first assumption means that failure will take place along the lines at which the condition for rupture is satisfied by way of yielding. The second assumption means that these lines will be straight. In the pure theory, curved yield lines occur. In the applied yield-line theory, these are replaced by polygons.

99. The applied yield-line theory further differs from the pure theory in that it does not require that the yield lines intersect each other in a definite way, i.e. at right-angles.

Conditions given by the applied yield-line theory

100. According to the conditions for rupture, the bending moment along the positive yield lines is constant and equal to m (the positive plastic moment) and the bending moment along the negative yield lines is constant and equal to $-m'$ (the negative plastic moment).

101. The yield lines are straight.

102. A yield line between two segments of a slab must pass through the intersection point of their axes of rotation.

103. The yield-line pattern is determined by the axis of rotation of the slab segments and the ratios of the rotations of the slab segments.

104. It has been assumed that the slab is equally reinforced both ways, so that m and m' are the same in all directions. If this is not so, the slab can, for the purpose of design, be transformed into one reinforced equally both ways by the use of affinity. The shear forces will then have to be modified.

SHEAR FORCES

105. Where yield lines of the same sign meet, the shear forces which are here concentrated into nodal forces are zero. See Fig. 15.

$$Q_A = Q_B = Q_C = 0$$

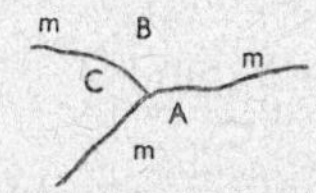

FIG. 15.—MEETING OF YIELD LINES OF SAME SIGN

106. Where yield lines of opposite sign meet, the shear forces or nodal forces are *not* zero. The nodal forces are as follows where two negative yield lines meet one positive (Fig. 16).

$$Q_A + Q_B + Q_C = 0, \text{ positive direction, out of the paper}$$
$$Q_A = (m+m') \cot \beta$$
$$Q_B = (m+m') \cot \alpha$$
$$Q_C = -(m+m') (\cot \alpha + \cot \beta)$$

107. Where two positive yield lines meet one negative, the shear forces are obtained by changing the signs of the shear forces above.

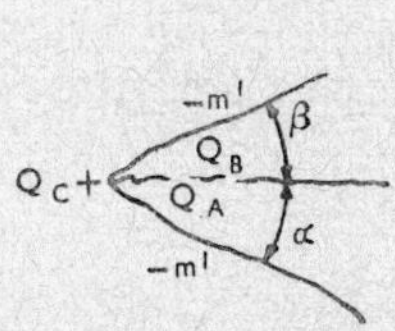

FIG. 16.—MEETING OF TWO NEGATIVE YIELD LINES AND ONE POSITIVE

108. Where a yield line meets a free edge, the shear forces are

$$Q_E = m \cot \alpha$$
$$Q'_E = -m \cot \alpha. \quad \text{(See Fig. 17.)}$$

109. Where yield lines of opposite sign meet, there is a maximum of three directions.

110. When more than one yield-line pattern is possible, the one which gives the maximum yield-line moment must be used.

Yield-line moments

111. The yield-line moments m and $-m'$ can be determined from the equations of equilibrium plus the geometrical conditions only.

112. In practice the principle of virtual work is often used. As this must be applied to a pre-determined yield-line pattern, the maximum condition must be introduced to find the correct yield-line pattern.

$$\sum m\,\theta\,l+\sum\int\int w\,z\,dx\,dy = 0$$

where θ is the angle of rotation of a slab segment, l the length of the yield line, w the load per unit area, and z the deflexion of the slab at any point.

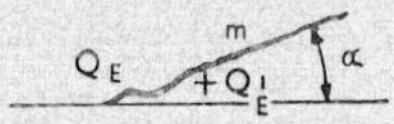

FIG. 17.—YIELD LINE MEETING A FREE EDGE

113. The possible formation of "corner levers" must be noted. If these "corner levers" are allowed to form, they have the effect of increasing the yield-line moment, thus weakening the slab.

Deflexions

114. Experience has shown that the check on deflexions in order to ensure that rotations are not excessive at any point before the complete yield-line pattern has formed is *not* normally necessary for slabs. Only a limited plastic zone is required to enable the full pattern to form. In the special case in which the behaviour in bending of a slab strip resembles that of a continuous beam, or when the assumed distribution of ultimate bending moments differs appreciably from that normally assumed for elastic conditions, the rotation of the hinges should be checked. In such cases, it is possible as in other plane structures for excessive plastic hinge rotations to be developed at ultimate load, as well as for undesirable cracking and deflexion to occur at working load.

115. Deflexions are only of interest at working load to check that they are within the permissible values. Maximum deflexion can be found by approximate elastic methods using the yield-line moment divided by the load factor as maximum moment, and applied to a strip cut out of the slab (only the order of magnitude is of interest).

"Shake-down"

116. The problem of "shake-down" can also occur for slabs, but has not so far been tackled theoretically. It is probably only of practical importance in extreme cases.

Tests

117. Tests to prove validity of the applied yield-line theory for slabs have been carried out by the Technical University of Denmark, by the Portland Cement Association in Chicago, U.S.A., and by the Cement and Concrete Association at Wexham Springs in England. The tests have generally shown good agreement between the test results and the calculated ones regarding both ultimate load and pattern of yield lines.

118. The plastic zone at the yield lines has been found to be fairly narrow compared with the depth of the slab.

Use up to the present time

119. Ultimate design for slabs has been used in practice to a great extent in the Scandinavian countries, France, U.S.A., and in Great Britain in recent years.

Prestressed concrete slabs

120. The yield-line theory for slabs has proved to be applicable also to slabs in prestressed concrete.[13, 23]

APPLICATION OF THEORY TO PRACTICAL DESIGN OF FRAMES

121. The calculation of θ values for common types of highly statically indeterminate systems such as building frames can be facilitated by the preparation of curves. Examples of such curves have been given by Tokarski[10] and Poologasundram[30]. The latter in particular gives values for three-bay frames with short intermediate spans. Baker[4] gives a worked example of the main sections of a building frame designed by ultimate load methods. Many existing structures have in effect been designed by ultimate load principles, but without regard for the influence of relative stiffness of members on permissible hinge rotations. For instance CP 114 has permitted, since 1939, the transfer of 15% of support moments to the adjacent spans. B.S.C.P. 115 allows for continuous beams and other statically indeterminate structures that the permissible compressive strength in the concrete may be increased from 0·33 u_w to 0·4 u_w within the range of support moments. Danish and Russian regulations also permit the transfer of bending moments.

122. Particular attention is drawn to Example I and its following rule for continuous beams of equal spans.

Practical application in Denmark

123. Ultimate load design is discussed in the Danish Code of Practice for Reinforced Concrete, D.S. 411, (preliminary issue Sept. 1944, published Nov. 1949). Clause 21 permits the design of statically indeterminate structures according to:

(*a*) elastic theory
(*b*) plastic theory
(*c*) the principle of "partial fixity".

124. Design according to the plastic theory simply requires that the internal forces are calculated according to the plastic properties of the materials. Alternatively it is sufficient if the internal forces are determined in such a way that the maximum internal forces in the structure at one point are not less than one third of the value derived from the elastic theory.

125. With simple structures (frames for houses, etc.) "partial fixity" can be used for the design of continuous beams and slabs. According to this principle, the bending moments at the supports are chosen between full fixity and one-fifth of that value.

126. The beam (or slab) is then designed for two bending moment diagrams:

(*a*) full load (live + dead) with two thirds of support moments;
(*b*) full support moments with dead + three fifths of live load.

127. The requirement that the internal forces be kept to not less than one

third of the values obtained by the elastic theory is an attempt to retain the serviceability of the structure. This is also the case for the requirements for "partial fixity" (support moment between one-fifth and full fixity and two-thirds of the support moment used in the diagram).

128. For the plastic design, it is assumed that working loads are used with "permissible stresses" and not ultimate load.

Practical application in Russia

129. *Russian recommendations for the design of continuous beams and slabs*[12, 24, 25]

(*a*) For slabs of equal span with ratio of sides l_2/l_1 greater than 2, the following formulae are recommended:

(i) $\frac{(W_d + W_e)}{16}l^2$ for intermediate span as well as support moments

(ii) $\frac{(W_d + W_e)}{11}l^2$ for end span moment

(iii) $\frac{(W_d + W_e)}{14}l^2$ for moment at supports second from end.

(*b*) For beams of equal span, the following formulae are recommended:

(i) $\frac{(W_d + W_e)}{16}l^2$ for intermediate span and support moments

(ii) $\frac{(W_d + W_e)}{11}l^2$ for span moments of end span as well as moment at penultimate support.

In the above formulae, $W_d + W_e$ are uniformly distributed dead and live loads respectively.

(*c*) For other continuous beams, it is recommended that design moments be assumed which are not less than 70% of the moments in the same sections calculated by the elastic theory. A section so designed should be so reinforced that the ratio of the neutral axis to the effective depth is less than or equal to 0·3.

Test results

130. Figs 2–8 in the Final Report[29] by Professor A. L. L. Baker on the meeting from 2 to 11 April, 1959 of the European Committee on Concrete show histograms of the ratio of measured bending strength to calculated bending strength for a variety of beam tests and confirm that the recommended values of parameters governing bending strength are sound.

131. The Figures relating to Tests Nos 3, 6, 10, 18, 19, and 36 in C.E.B. Bulletin No. 21[7] show moment/rotation curves obtained from tests compared with the limits L_1 and L_2 for the beam sections, and confirm that the parameter values governing the limits L_1 and L_2 are safe and reasonable. Further tests are being carried out in various parts of the world in an international programme co-ordinated by the European Committee on Concrete.

Co-ordination of research on inelastic deformation

132. When Commission XI of the European Committee on Concrete met in Stockholm in June 1960, it was agreed that the various directors of laboratories who were interested in carrying out tests to determine moment/rotation characteristics should be asked to collaborate by producing their results in a standard

way so that the results could be co-ordinated without difficulty. The results would be published in a bulletin.

133. At a later meeting at Monaco in January 1961 some preliminary test reports were received and the work of carrying out tests and studying the influence of each of the parameters affecting moment/rotation characteristics was finally divided among 12 laboratories in Europe and U.S.A.

134. Other items referred to in this report requiring research are:

(*a*) Permissible values of θ

(*b*) Influence of plastic hinges on shear strength.

WORKED EXAMPLES

EXAMPLE 1: CONTINUOUS BEAM

135. Let xM be the bending moment at any support section of a continuous beam of at least five equal spans, except the end and penultimate supports. The free moment diagram is parabolic, and has maximum value M. Details of the beams are assumed to have been worked out as follows:—

$M = 0{\cdot}1\ bd^2c_u$

$E = 1{,}000c_u$ (a high value since beam is unuer-reinforced)

$I = 0{\cdot}05bd^3$

$l = 10d$

$z = 0{\cdot}15l$, so that $\frac{z}{d} = 1{\cdot}5$

$k_1 k_3 = 0{\cdot}5$

$k_2 = 1$

$\frac{P}{P_u} = 0$

$n_u d = 0{\cdot}2d$

$e_{cu} - e_{ce} = 0{\cdot}002$ (a high value since beam is under-reinforced and e_{ce} assumed$=0{\cdot}0015$

From equation (5), assuming plastic hinges to occur at the support sections:

$$-\frac{2}{3}\frac{Ml}{EI} + x\frac{Ml}{EI} = -\theta \qquad \text{(note when } \theta = 0,\ x = \tfrac{2}{3}\text{)}$$

Suppose it is required to make $x = 0{\cdot}5$

Then $\theta = 0{\cdot}16\frac{Ml}{EI}$ Substituting for M, E and I,

$$\theta = 0{\cdot}00032\frac{l}{d} \qquad \left(\text{note that } \theta \text{ varies with } \frac{l}{d}\right)$$

when $\frac{l}{d} = 10$. $\theta = 0{\cdot}0032$

Permissible $\theta_p = \frac{0{\cdot}002}{0{\cdot}2d} . k_1 k_3 \left(\frac{z}{d}\right)^{0{\cdot}25}$. d multiplied by 2 for both sides of critical section.

Substituting the above values, permissible $\theta_p = 0{\cdot}01$

The sign of θ is positive. Plastic hinges therefore occur at the supports as assumed.

136. If the beam is designed for support bending moments of value $\frac{M}{2}$, the rotations of the plastic hinges will not be excessive. There is no need, therefore, to use closed binders in order to increase the permissible value of $e_{cu} - e_{ce}$, which would be necessary for higher values of $\frac{l}{d}$ and $n_u d$, or lower values of x.

Rule for equal-span continuous beams

137. In continuous beams of approximately equal span supporting uniformly distributed load the support moments (except penultimate) may be assumed equal to the mid-span moments provided the permissible value of θ for the support sections is greater than $0{\cdot}16\frac{Ml}{EI}$.

EXAMPLE 2

138. The following is an example worked out using published curves[30] by Poolagasundram (to which reference must be made) to illustrate a possible design method by the use of curves. These curves make certain assumptions which may not apply to all practical cases. However, other curves can be drawn in the design office to widen their application.

139. The design of the rth storey of a three bay frame with external beams spanning 16 ft 8 in. and the internal beam spanning 10 ft 10 in. is illustrated. The following design data with a load factor of 2 are assumed:—

(*a*) All columns 12 ft 0 in. high with sizes 15 in. × 12 in.
(*b*) The free B.M. (M_1) of the external beams at ultimate load is 2,000,000 lb-in.
(*c*) The free B.M. (M_2) of the internal beam at ultimate load is 450,000 lb-in.
(*d*) The free B.M. (m) due to wind is 2,500,000 lb-in.
(*e*) Direct load on all columns at ultimate is 550,000 lb.
(*f*) Beams have section of $b=9$ in. and $d=16$ in.

140. The following constants are easily determined for using the curves:—

$$y = 0{\cdot}6900$$
$$\frac{M_1 l}{EI} = 0{\cdot}0625$$
$$r_M = 0{\cdot}2250$$
$$r_m = 1{\cdot}2500$$
$$K = 0{\cdot}6000$$

TABLE 6.—HINGE ROTATIONS

Hinge	Y_{M_1}	Y_{M_2}	Y_m	$\theta = \frac{M_1 l}{EI}(Y_{M_1} + r_M Y_{M_2} + r_m Y_m)$	θ
1-*r*-*b*	−0·062	0·200	0·020	0·0625 (−0·062+0·045+0·025)	0·0005
2-*r*-*b*	−0·070	0·200	−0·006	,, (−0·070+0·045−0·007)	−0·0020
3-*r*-*b*	0·127		0·022	,, (0·127+0+0·027)	0·0096
2-*r*-*a*	0·217	−0·200	0·065	,, (0·217−0·045+0·081)	0·0158
3-*r*-*a*	−0·014		0·065	,, (−0·014+0+0·081)	0·0042
1-*r*-*c*	−0·020		0·020	,, (−0·020+0+0·025)	0·0003
2-*r*-*c*	0·217	−0·200	0·060	,, (0·217−0·045+0·075)	0·0155
3-*r*-*c*	−0·014		0·070	,, (−0·014+0+0·087)	0·0046
4-*r*-*c*	0·020		0·007	,, (0·020+0+0·009)	0·0018

141. Assuming $\theta_{2\text{-r-a}}$ and $\theta_{2\text{-r-c}}$ are excessive and the permissible values are only 0·0100, these θ values are reduced by increasing $X_{2\text{-r-a}}$ and $X_{2\text{-r-c}}$ by 112,000 lb-in. The approximate increment of hinge moment to effect this can be obtained by dividing the excess of the rotation by the influence coefficient. Table 7 gives the resultant hinge rotations due to the increase of these hinge moments. $\theta_{1\text{-r-c}}$ although negative is the correct sign because the final $X_{1\text{-r-c}}$ value is of the opposite sign to the assumed one. $\theta_{1\text{-r-b}}$ is considered to be near enough to zero.

TABLE 7.—RESULTANT θ VALUES

Hinge	Change in θ due to $X_{2\text{-r-a}}$	Change in θ due to $X_{2\text{-r-c}}$	Final θ
1-*r*-*b*	0·0001	0·0001	0·0007
2-*r*-*b*	0·0007	0·0007	−0·0006
3-*r*-*b*	−0·0020	−0·0020	0·0056
2-*r*-*a*	−0·0035	−0·0023	0·0100
3-*r*-*a*	−0·0020	−0·0020	0·0002
1-*r*-*c*	−0·0012	−0·0012	−0·0021
2-*r*-*c*	−0·0023	−0·0035	0·0100
3-*r*-*c*	−0·0020	−0·0020	0·0006
4-*r*-*c*	0	0	0·0018

142. The final hinge moments are as follows:—

$\bar{X}_{1\text{-r-b}}$ = 1,325,000 lb-in. $\bar{X}_{1\text{-r-c}}$ = 122,000 lb-in.
$\bar{X}_{2\text{-r-b}}$ = 1,325,000 lb-in. $\bar{X}_{2\text{-r-c}}$ = 437,000 lb-in.
$\bar{X}_{3\text{-r-b}}$ = 1,400,000 lb-in. $\bar{X}_{3\text{-r-c}}$ = 310,000 lb-in.
$\bar{X}_{2\text{-r-a}}$ = 437,000 lb-in. $\bar{X}_{4\text{-r-c}}$ = 622,000 lb-in.
$\bar{X}_{3\text{-r-a}}$ = 310,000 lb-in.

EXAMPLE 3

143. The following example serves only to illustrate the design procedure discussed in the report. Experienced designers may find it unnecessary to go into such details.

144. Fig. 18 gives the dimensions of a three-bay reinforced concrete frame for a single-storey factory building with precast roof units.

145. The ultimate external loadings the frame is assumed to carry are shown in Fig. 19 assuming a load factor of 2.

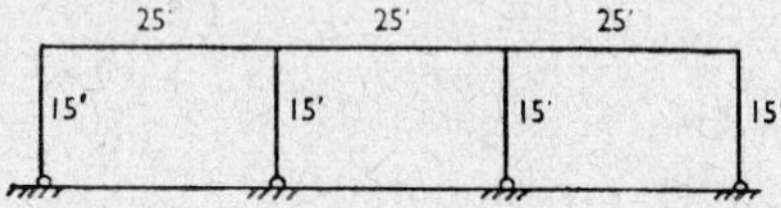

FIG. 18

146. The appropriate beam and column sections are assumed to be 27 in. × 12 in. and 18 in. × 12 in. respectively. Using the safe limiting values of parameters recommended in this report and following procedure as given in reference 4, the following *EI* values are arrived at by trial and error.

$$EI \text{ for beam} = 40{,}000 \times 10^6 \text{ lb-in.}^2$$
$$EI \text{ for column} = 12{,}560 \times 10^6 \text{ lb-in.}^2$$

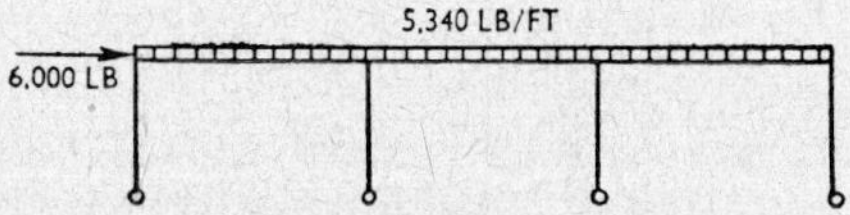

FIG. 19

147. Two cases of loading are considered; case 1 when only uniformly distributed loads are acting and case 2 when horizontal and vertical loads are acting simultaneously.

148. *Case* 1. For this condition of loading, hinges are assumed at the positions given in Fig. 20. This system of hinges does not form a mechanism for symmetrical load. Although there are six hinges, only three rotations need to be evaluated due to symmetry. The hinges are identified as b_1, b_2 and b_3 as shown in the diagram. To

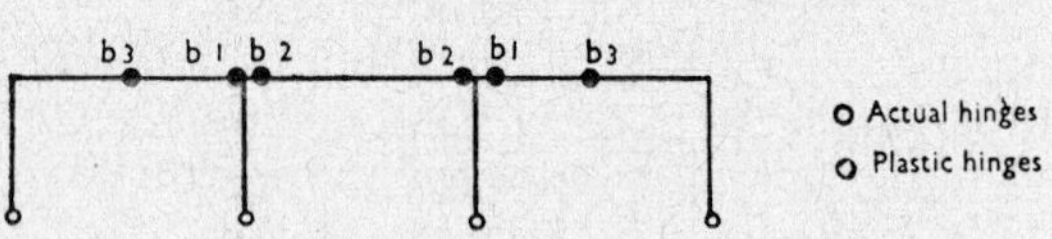

FIG. 20

work out the influence coefficients, unit moments are applied at the various hinges and the resultant moment diagrams are given in Fig. 21.

149. The influence coefficients (δ_{ik}) obtained by integrating these moment diagrams are listed in Table 8.

150. The moment diagram when external load is acting on the system is shown in Fig. 22 while the corresponding influence coefficients are given in Table 9.

151. The expression giving hinge rotations is as follows:—

$$-\theta_{1b} = X_{1b}\,\delta_{1b\cdot 1b} + X_2\,\delta_{1b\cdot 2b} + X_3\,\delta_{1b\cdot 3b} + \delta_{01b}$$

152. The values of hinge moments are adopted from the following considerations:—

The free moment due to uniformly distributed load of 5,340 lb per ft =

$$5{,}340 \times 25 \times 25 \times 1{\cdot}5 \text{ lb-in.} = 5{\cdot}0 \times 10^6 \text{ lb-in.}$$

It would be ideal to choose $X_{1b} = X_{2b} = X_{3b} = \frac{1}{2} \times 5{\cdot}0 \times 10^6$ lb-in.

TABLE 8

Influence coefficients, δ_{ik}	$1b$	$2b$	$3b$
$1b$	$0{\cdot}02412\times10^{-6}$	$-0{\cdot}00956\times10^{-6}$	$0{\cdot}0225\times10^{-6}$
$2b$	$-0{\cdot}00956\times10^{-6}$	$0{\cdot}01700\times10^{-6}$	0
$3b$	$0{\cdot}0225\times10^{-6}$	0	$0{\cdot}0583\times10^{-6}$

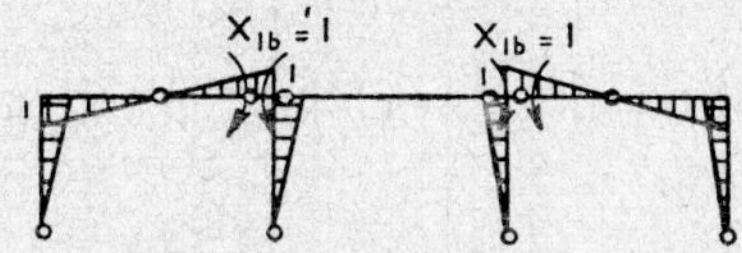

Condition M_1(unit moment applied at hinges b_1)

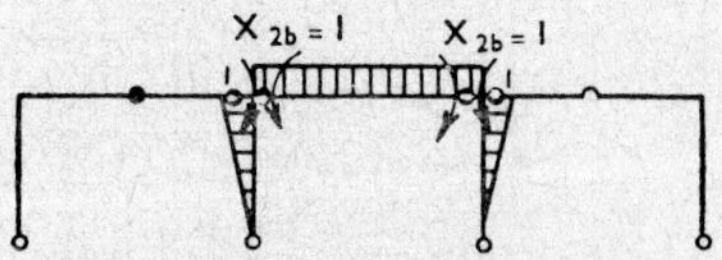

Condition M_2(unit moment applied at hinges b_2)

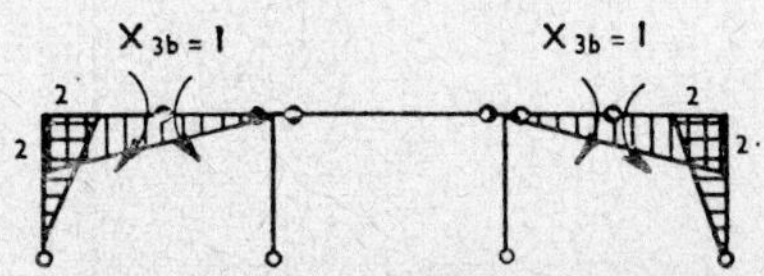

Condition M_3(unit moment applied at hinges b_3)

FIG. 21

TABLE 9

δ_{01b}	δ_{02b}	δ_{03b}
$-0{\cdot}12036$	$-0{\cdot}02500$	$-0{\cdot}24142$

153. However, by so doing, the external column will have to take a moment of $2{\cdot}5\times10^6$ lb-in. It may be desirable to reduce the moment in the column and a higher value such as $2{\cdot}9\times10^6$ lb.-in. is used for $\bar{X}_{3b}$ and $\bar{X}_{1b}$. To accommodate additional moment due to the action of horizontal load in Case 2, instead of $2{\cdot}5\times10^6$ lb-in. assume $X_{2b}=2{\cdot}7\times10^6$ lb-in. With these trial values of $\bar{X}_{1b}$, $\bar{X}_{2b}$ and $\bar{X}_{3b}$, the hinge rotations are tabulated in Table 10.

TABLE 10

	$\bar{X}$	1b		2b		3b	
		δik	$\bar{X}\delta ik$	δik	$\bar{X}\delta ik$	δik	$\bar{X}\delta ik$
1b	$2 \cdot 9 \times 10^6$	$0 \cdot 02412 \times 10^{-6}$	0·07000	$-0 \cdot 00956 \times 10^{-6}$	−0·027800	$0 \cdot 0225 \times 10^{-6}$	0·065300
2b	$2 \cdot 7 \times 10^6$	$-0 \cdot 00956 \times 10^{-6}$	−0·025850	$0 \cdot 01700 \times 10^{-6}$	0·045900	0	0
3b	$2 \cdot 9 \times 10^6$	$0 \cdot 02250 \times 10^{-6}$	0·065300	0	0	$0 \cdot 0583 \times 10^{-6}$	0·169300
δ_0			−0·120360		−0·02500		−0·24142
$-\theta$			−0·010910		−0·00690		−0·00682

154. It is seen from Table 10 that the moment distribution is possible if hinge rotations are not excessive. To check this, consider hinge $1b$.

Referring to equation (8) in the text,

Assume in this case $k_1 = 0{\cdot}7$, $k_2 = 1$ and $k_3 = 0{\cdot}8$, $z = 0{\cdot}25l$

$$n_u = 0{\cdot}18$$

$$\theta_p = \frac{0{\cdot}0015}{0{\cdot}18} \times 0{\cdot}735 = 0{\cdot}0061$$

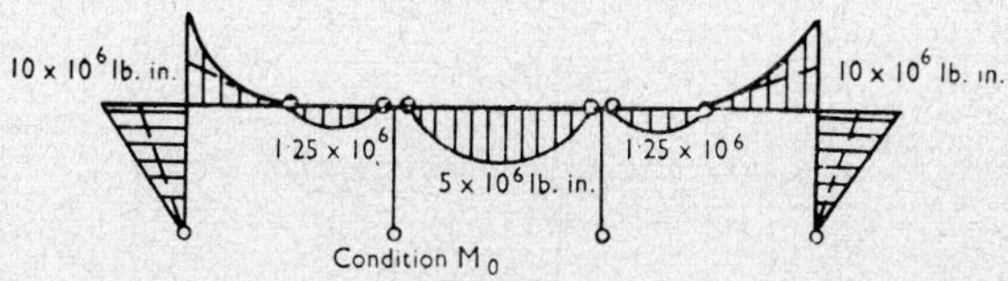

FIG. 22

155. It is clear that the rotation is excessive and the plastic moment at hinges $1b$ need to be increased. Examining the rotation of hinge $2b$ in Table 10, it is seen that it will be advisable to increase $X2b$ simultaneously to avoid further increase in $\theta 2b$. Table 11 gives the adjustments of moments and the resultant alteration in rotations.

156. The resultant rotations of the various hinges as given in Table 11 can be considered satisfactory. The moment distribution for this system is given in Fig. 23.

157. *Case* 2. When both horizontal and vertical loads are acting. For this condition of loading hinges are assumed at positions as shown in Fig. 24.

TABLE 11

	$\Delta \bar{X}$	$\Delta \bar{X}.\delta ik$		
		$1b$	$2b$	$3b$
$1b$	$0{\cdot}3 \times 10^6$	0·00725	−0·00287	0·006750
$2b$	$0{\cdot}2 \times 10^6$	−0·00191	0·00340	0
$3b$	0	0	0	0
$\Delta\theta$		0·00534	0·00053	0·00675
Original $-\theta$		−0·010910	−0·00690	−0·00682
Resultant $-\theta$		−0·00557	−0·00637	−0·00007

158. The moment diagrams when unit moment is applied at the various hinges are given in Fig. 25.

159. When horizontal and vertical loads are applied to this system, the resultant moment diagrams are given in Fig. 26.

160. The influence coefficients obtained by integrating these moment diagrams are tabulated in Table 12. As in previous cases, the expression giving hinge rotations is as follows:—

$$-\theta_{1b} = \bar{X}_{1b}\,\delta_{1b\cdot 1b} + \bar{X}_{2b}\,\delta_{2b\cdot 1b} + \bar{X}_{3b}.\delta_{3b\cdot 1b} + \bar{X}_{1a}.\delta_{1a\cdot 1b} + \bar{X}_{2a}.\delta_{2a\cdot 1b} + \delta_{1b\cdot OM} + \delta_{1b\cdot om}$$

. . . . etc. . . . . .

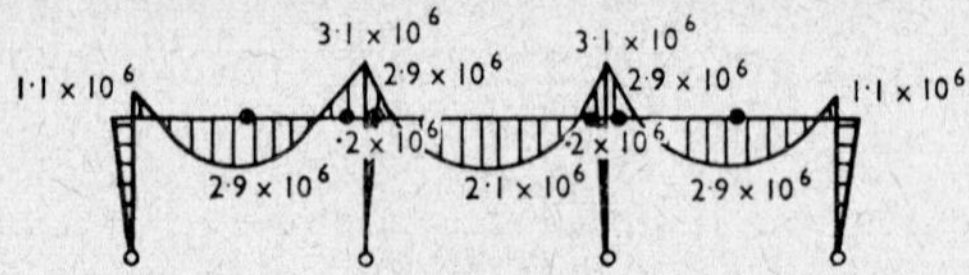

FIG. 23

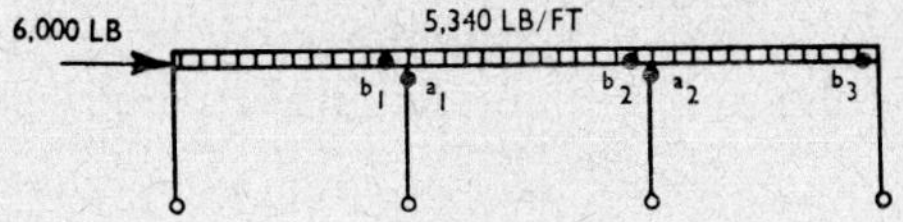

FIG. 24

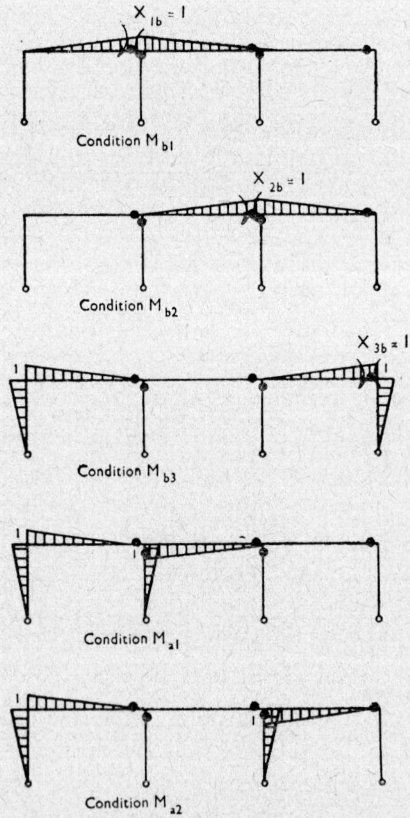

FIG. 25

TABLE 12

	1b	2b	3b	1a	2a
1b	$0{\cdot}00500\times10^{-6}$	$0{\cdot}00125\times10^{-6}$	$0{\cdot}00125\times10^{-6}$	$-0{\cdot}00125\times10^{-6}$	$0{\cdot}00125\times10^{-6}$
2b	$0{\cdot}00125\times10^{-6}$	$0{\cdot}00500\times10^{-6}$	$0{\cdot}00125\times10^{-6}$	$-0{\cdot}00125\times10^{-6}$	$-0{\cdot}00250\times10^{-6}$
3b	$0{\cdot}00125\times10^{-6}$	$0{\cdot}00125\times10^{-6}$	$0{\cdot}01450\times10^{-6}$	$0{\cdot}00730\times10^{-6}$	$0{\cdot}00605\times10^{-6}$
1a	$-0{\cdot}00125\times10^{-6}$	$-0{\cdot}00125\times10^{-6}$	$0{\cdot}00730\times10^{-6}$	$0{\cdot}01450\times10^{-6}$	$0{\cdot}00730\times10^{-6}$
2a	$0{\cdot}00125\times10^{-6}$	$-0{\cdot}00250\times10^{-6}$	$0{\cdot}00605\times10^{-6}$	$0{\cdot}00730\times10^{-6}$	$0{\cdot}01450\times10^{-6}$
oM	$-0{\cdot}02500$	$-0{\cdot}02500$	$-0{\cdot}02500$	0	0
om	$-0{\cdot}00135$	0	$-0{\cdot}00785$	$-0{\cdot}00785$	$-0{\cdot}00785$

161. Referring to Fig. 23, the hinge moments as a first trial should be as follows:—

$$\bar{X}1b = 3{\cdot}1\times10^6$$
$$\bar{X}2b = 2{\cdot}9\times10^6$$
$$\bar{X}3b = 1{\cdot}1\times10^6$$
$$\bar{X}1a = 0{\cdot}2\times10^6$$
$$\bar{X}2a = 0{\cdot}2\times10^6$$

The evaluation of the hinge rotations is listed in Table 13.

TABLE 13

	1*b*	2*b*	3*b*	1*a*	2*a*
$-\theta$	−0·005845	−0·005995	−0·006740	−0·002960	−0·000210

162. The hinge rotations are seen to be satisfactory and the resultant moment distribution at ultimate is given in Fig. 27.

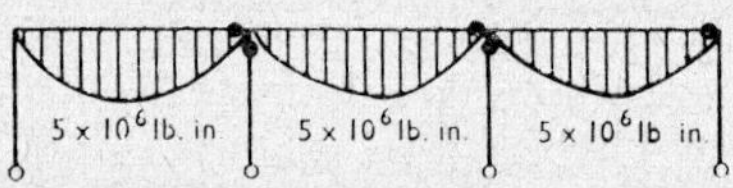

Condition M_{OM}

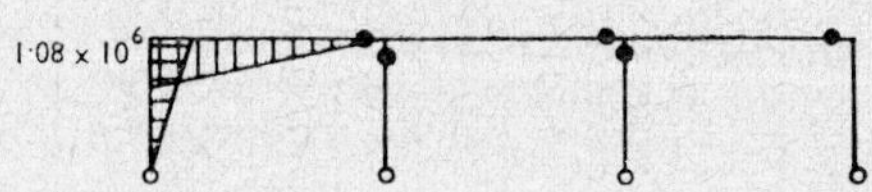

Condition M_{OM}

FIG. 26

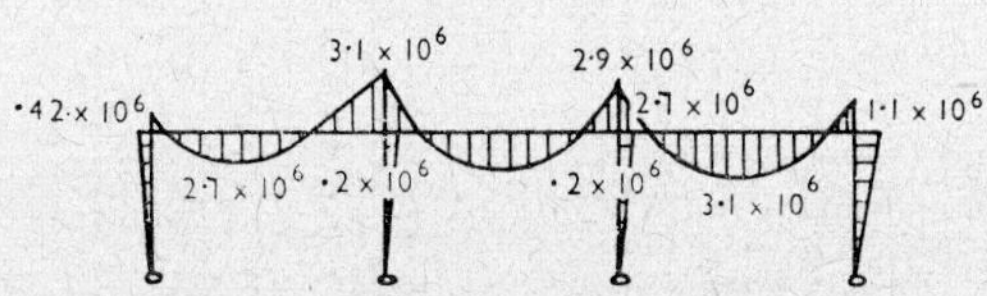

FIG. 27

163. It should be noted that according to the moment distribution given in Fig. 27, the mid-span section of the end span remote from the side where lateral force is assumed to act should be reinforced for $3{\cdot}1\times10^6$ lb-in. In case 1, a hinge is assumed at this section and a hinge moment of $2{\cdot}9\times10^6$ lb-in. is used. Strictly speaking, in order to satisfy both conditions of loading, the hinge moment at 3*b* of Case 1 should be $3{\cdot}1\times10^6$ lb-in. and the procedure for Case 1 repeated again. However, in this case the difference is so small and also the rotations for 3*b* in Case 1 is almost zero indicating that the hinge is not yielding very much, it should therefore, be satisfactory to reinforce that section for $3{\cdot}1\times10^6$ lb-in. and assume that a slight increment of load beyond 5,340 lb per ft will produce a hinge at 3*b*. For cases where the lateral load predominates, the case for lateral load acting alone should be attempted first.

REFERENCES

1. R. Gartner, "Statically-indeterminate structures". 3rd ed., 1958, Concrete Publications, London.
2. A. L. L. Baker, "Reinforced concrete". 1949, Concrete Publications, London.

3. G. Macchi, "Redistribution of moments in inelastic field and rupture in prestressed concrete beams". Proceedings of Symposium on Plasticity in the Science of Construction, Varenna 1956, pub. N. Zanichelli, Bologna, 1957.
4. A. L. L. Baker, "The ultimate load theory applied to the design of reinforced and prestressed concrete frames". 1956, Concrete Publications, London.
5. H. Rüsch, "Researches toward a general flexural theory for structural concrete" J. Amer. Conc. Inst., July, 1960.
6. Y. Guyon. "The strength of statically indeterminate prestressed concrete structures", Proc. Symposium on the Strength of Concrete Structures, May 1956. Cem. Conc. Assn, London, pp. 305–376.
7. Comité Européen du Béton, Bulletin d'Information No. 21, note and appendices, 1960, Paris.
8. W. Hansell and G. Winter, "Lateral stability of reinforced concrete beams". J. Amer. Conc. Inst., Sept. 1959, pp. 193–213.
9. R. H. Wood, "The stability of tall buildings". Proc. Instn civ. Engrs, Sept. 1958, pp. 69–102 and 524.
10. E. W. Tokarski, "Application of the plastic-hinge method to a multi-storey frame". Conc. and Const. Engg, Feb. 1959, pp. 85–95.
11. R. Gartner, "Design of statically indeterminate structures by the plastic hinge method". Conc. and Const. Engg, May, 1957, pp. 161–174.
12. K. E. Tal, "Limit design of concrete and reinforced concrete structures". 1955, State Publications, Moscow.
13. Aa. Ingerslev, "Om en elementaer Beregningsmaade af krydsarmerede Plader". Ingeniøren, Aug. 1921. Copenhagen, pp. 507–515.
14. Aa. Ingerslev, "The strength of rectanguler slabs". Struct. Engr, January 1923, London, pp. 3–19.
15. O. Graf, "Versuche mit allseitig aufliegenden rechteckigen Eisenbeton-Platten unter gleichmässig verteilter Belastung". Bulletin No. 56, 1926. Deutscher Ausschuss für Eisenbeton, Berlin.
16. C. Bach, "Versuche mit zweiseitig aufliegenden Eisenbetonplatten bei konzentrierter Belastung", Part 2, Bull. No. 52, 1923, Deutscher Ausschuss für Eisenbeton, Berlin.
17. K. W. Johansen, "Beregning af krydsarmerede Jaernbetonpladers Brudmoment". 1931, Bygningsstatiske Meddelelser, Copenhagen.
18. K. W. Johansen, "Nogle Pladeformler". 1932, Bygningsstatiske Meddelelser, Copenhagen.
19. K. W. Johansen, "Brudlinieteorier". July 1943, Gjellerups Forlag, Copenhagen.
20. K. W. Johansen, "The ultimate strength of reinforced concrete slabs". Final Report, Third Congress, Internat. Assn for Bridge and Struc. Engg, Sept. 1948, Liège, pp. 565–570.
21. K. W. Johansen, "Bruchmomente der kreuzweise bewehrten Platten". Publications, vol. 1, International Assn for Bridge and Struct. Engng, Zurich, 1932, pp. 276–296.
22. K. W. Johansen, "Pladeformler". 2 Udgave, Polyteknisk Forening, Copenhagen, 1949.
23. K. W. Johansen, "Pladeformler; Formelsamling". 2 Udgave, Polyteknisk Forening, 1949, Copenhagen.
24. C. W. Yu, M. Corbin and E. Hognestad, "Reinforced concrete design in the U.S.S.R." J. Amer. Conc. Inst., vol. 31, No. 1, July 1959, pp. 65–69.
25. Ministry of Industrial Construction (U.S.S.R.), "Instruction for the design of sections of reinforced concrete structural elements". State Publishing Office of Literature for Structural Engineering and Architecture, Moscow, 1956. (In Russian).
26. C. W. Yu, and E. Hognestad, "Review of limit design for structural concrete". J. Struct. Div., Amer. Soc. civ. Engrs, Dec. 1958, Paper 1878, pp. 1–28.

27. C. W. Yu, "An analytical investigation of the ultimate load design of reinforced concrete framed structures". Univ. of London Ph.D. thesis, 1954.

28. A. L. L. Baker, "A general equation for frame analysis". Conc. Const. Engng, March 1961, pp. 97–103.

29. A. L. L. Baker, "European Committee on Concrete". Reinf. Conc. Rev., Dec. 1959, pp. 247–264.

30. K. Poologasundram, "The analysis of three-bay frames by the plastic hinge method" Conc. Const. Engng, vol. 56, Sept. 1961, pp. 301–306.

31. W. W. L. Chan, "The ultimate strength and deformation of plastic hinges in reinforced concrete frameworks". Mag. of Conc. Res., Cem. Conc. Assn, No. 21, Nov. 1955, pp. 121–132.

32. K. Poologasundram, "An analytical and experimental investigation of the formation and behaviour of plastic hinges in prestressed and reinforced concrete frames". Univ. of London Ph.D. thesis, 1959.

33. A. L. L. Baker, "A general analysis of elasto-plasitic 3-dimensional frames." Internat. Assocn. Bridge & Struct. Engg., Zurich, 1960.

WRITTEN DISCUSSION

Mr. A. N. Procter (Consulting Engineer, Nottingham) wrote that stress distributions similar to those illustrated in Fig. 8b might be assumed in the concrete slabs and casings of composite construction to provide for normal inelasticity, creep and shrinkage. Formulas based on this distribution applied not only to ultimate load conditions, but gave remarkably accurate results at working loads. Test data published by Professors Mackay, Gillespie, and Lelnau[34] might be used as an example.

165. A reinforced concrete slab 4-in. thick and 10 ft wide was cast integral with two American 10-in. standard I-beams 22·6 lb/ft, 5 ft apart and 16 ft long (Fig. 28). The

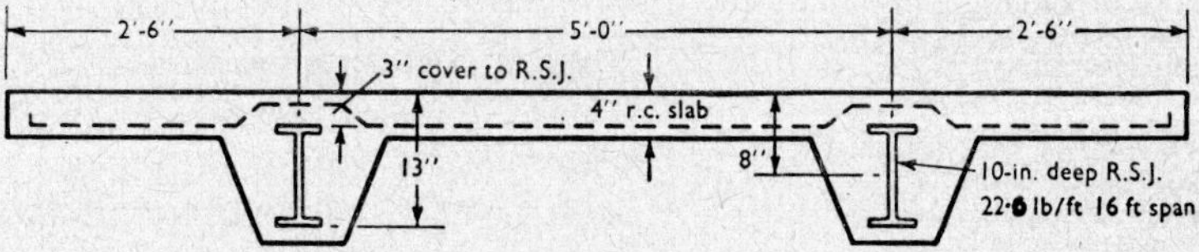

FIG. 28

beams were encased in concrete and had 3 in. top cover. In three tests the maximum bending moments of 460, 565 and 845 kip-inches induced measured maximum steel stresses of 11,810, 14,350 and 21,500 lb/sq. in. respectively. Assuming a modular ratio of 18, the depth of neutral axis was 4 in. by elastic theory. The stresses at the top of the concrete slab were then calculated to be 300, 365 and 536 lb/sq. in. respectively. Assuming the effective width of slab to be 58 in., the corresponding bending moments by elastic theory were 377, 458 and 684 kip-inches, about 20% lower than the applied bending moments.

166. Using formulas (10), § 168, and 11, § 171, for the same stresses in the steel and concrete the bending moments were respectively 466, 567 and 848 kip-inches (from $d_1 = 13$ in., $d_2 = 10$ in., $d_3 = 4$ in., $b_1 = 8$ in. and $b_2 = 48$ in.). These calculated values were almost exactly the same as the test figures. The calculated mean bond stresses were 48, 59, and 88 lb/sq. in., so there was probably no slip. Similar examples might be quoted, showing that the ultimate load method might also be used to represent working conditions accurately. It was found that the appropriate formulas (10) to (12) might be used to predict with reasonable accuracy the ultimate strength of composite beams,[35] provided that slip between concrete and steel did not occur before failure. Adequate allowances for fatigue, referred to in § 68(d) of the report, were of particular importance, since the action of shear connectors was modified profoundly by repeated loading and fatigue. The derivation of formulas (10) to (12) is given below.

167. Figs 29a and b represented composite beams of rolled steel I and split I-section encased in structural concrete. d_1 was the depth of the underface and d_2 was the depth

of the neutral axis of the steel below the concrete, b_1 being the effective width of the concrete casing. Considering first an encased I-beam, the elastic neutral axis was found to be just about $\frac{1}{2}d_1$ below the top of concrete. To allow for normal inelasticity, creep and shrinkage, the stress distribution in the concrete was represented diagramatically by a parabola, as shown in Fig. 29*a*.

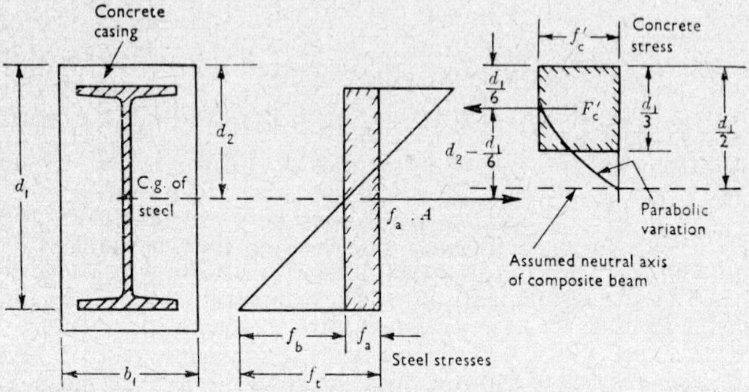

FIG. 29A.—ENCASED BEAM

168. For a maximum compressive stress f'_c the compressive force F'_c in the concrete was approximately $\frac{2}{3}f'_c b_1\frac{d_1}{2}=\frac{1}{3}f'_c b_1 d_1$. The centre of pressure was $\frac{d_1}{6}$ below the top of the concrete, so that F'_c might be represented by a rectangle of depth $\frac{d_1}{3}$. In the steel there was an equal and opposite force, giving a tensile axial stress on an area, A, of $f_a=f'_c b_1 d_1/3A$. The lever arm was $d_2-\frac{d_1}{6}$. The flexure of the concrete was ignored and the area of concrete displaced by the steel was not deducted. If there was a maximum bending stress of f_b in the steel, the maximum tensile stress would be $f_t=f_a+f_b$. Thus the total internal moment, $M_R=f_b Z_x+\frac{1}{3}f'_c b_1 d_1\left(d_2-\frac{d_1}{6}\right)$, where Z_x was the section modulus of the I-beam about its neutral axis. Since $f_b=f_t-f_a$, the moment of resistance

$$M_R = Z_x\left(f_t-\frac{1}{3}f'_c b_1 d_1/A\right)+\frac{1}{3}f'_c b_1 d_1\left(d_2-\frac{d_1}{6}\right).$$

$$Z_x/A = z_1, \text{ the major core radius,}$$

$$\text{thus } M_R = f_t Z_x+\frac{1}{3}f'_c b_1 d_1\left(d_2-\frac{d_1}{6}-z_1\right) \quad . \quad . \quad . \quad . \quad . \quad . \quad . \quad (10)$$

169. The bond stress between the steel and concrete casing was caused by the rate of change of the concrete force, F'_c. The underside of the I-beam was ineffective, so that the steel perimeter bonding with the casing was assumed to be $o=2\times\text{depth}+3\times\text{the}$ breadth of the I-beam. For a central point load on a span l the mean bond stress would be $\frac{2f'_c b_1 d_1}{3lo}$. Under uniformly distributed loading the mean bond stress near the supports was $\frac{4f'_c b_1 d_1}{3lo}$. It was found experimentally that the permissible bond stresses on I-beams was rather less than was used for reinforced concrete. The bond might however be increased by means of shear connectors similar to those used in tee-beams. (§ 175)

170. The depth of the neutral axis of an encased split I- or tee-section was about

$\frac{3}{4}d_2$. The lever arm would then be $\frac{5}{8}d_2$ and $F'_c = f'_c b_1 d_2/2$. The moment of resistance,

$$M_R = f_t Z_x + \tfrac{1}{2} f'_c b_1 d_2 (\tfrac{5}{8} d_2 - z_1) \quad . \quad . \quad . \quad . \quad . \quad . \quad . \quad (10a)$$

The bond stress for a central point load was $f'_c b_1 d_2/lo$ and $2f'_c b_1 d_2/lo$ for uniformly distributed load.

171. Tee-beams consisted essentially of concrete slabs bonded to steel beams. Sometimes, however, the steel was encased in concrete. When the thickness of the slab, d_3, was less than $d_1/3$ the compression in a fully stressed slab of width b_2 was very nearly $f'_c b_2 d_3 = F'_c$. The axial stress in the steel was $f'_c b_2 d_3/A$, so that

$$M_R = f_t Z_x + f'_c b_2 d_3 (\tfrac{1}{2} d_1 - z_1) \quad . \quad . \quad . \quad . \quad . \quad . \quad . \quad . \quad (11)$$

Shear connectors were required for the bond and under a central point load the shear per unit length of beam would be $2f'_c b_2 d_3/l$. With uniformly distributed loading the maximum shear per unit length was $4f'_c b_2 d_3/l$ at the ends. Tests showed that propping during construction did not increase the ultimate strength[35].

172. When the width of slab was greater than the critical width, $b_3 = \dfrac{3f_t A}{2f'_c d_3}$, the steel became overstressed. Substituting b_3 for b_2 in formula (11),

the critical moment of resistance, $M_c = f_t A(\tfrac{3}{4} d_1 - \tfrac{1}{2} z_1)$ . . . . (12)

The maximum horizontal shear with mid-span loading would then be $3f_t A/l$ per unit length, and $6f_t A/l$ per unit length at the supports for uniformly distributed loading. Camber might be required to allow for shrinkage.

173. When the depth of the slab was less than $d_1/3$, formula (10) might be used for a fully stressed slab. If, however, the width exceeded the critical width, $b_3 = 4{\cdot}5 f_t A/f'_c d_1$, the critical moment of resistance of the beam would be $M_c = f_t A(1{\cdot}5 d_2 - \tfrac{1}{4} d_1 - \tfrac{1}{2} z_1)$. The shear per unit length would still be $3f_t A/l$ for mid-span loading and $6f_t A/l$ for uniformly distributed loading.

174. The additional strength of a concrete casing might be included in the moment of resistance of a tee-beam by adding the term $\dfrac{1}{3} f'_c b_1 d_1 \left(d_2 - \dfrac{d_1}{6} - z_1 \right)$ to formula (11). When, however, the critical width was exceeded, the casing might be ignored, because the steel was then fully stressed, but it improved the bond.

175. The bond between the concrete and steel might be made with connectors and anchors that absorbed the diagonal tension. Stud and channel connectors were designed by empirical rules, published elsewhere[35, 35a]. Their spacing might be found from the horizontal shear formulas given above and they were used to prevent the slab from lifting. Anchors distributed the force through the width of the slab (Fig. 29*b*) and might be designed by the rules of reinforced concrete. About $\frac{3}{4}F'_c$ should be absorbed in the outer thirds of a span. The remainder was provided for in both directions in the middle third to allow for reversals. Alternatively the slabs might be precast and bolted to the steelwork with high-tensile friction-grip bolts. Connectors had no endurance limit, so their load factor should be substantial. Further test data was required before high tensile steel ribs could be included.

Mr H. L. Su (Lecturer in Civil Engineering, Chalmers University of Technology, Göteborg, Sweden), wrote that the inconsistency of the limit design philosophy, on which the ultimate load method was based, had been revealed recently[36–40]. Methods described in the Report, as a consequence, contained many loopholes. A detailed anatomy of such methods was forthcoming,[40] so his discussion would be focused upon a few major points in the Report. Because of the restricted space, many important topics, such as the stress/strain law, instability, deformation, etc., had been omitted.

177. It was interesting to note that the title "plastic design", originally used to designate the same ultimate load method,[41, 42] was discarded in favour of the precent name, presumably after an incorrect criticism[43] from a "steel plastician".[43] Ironically, if the analysis based on the so-called plastic deformation theory, which was the plastic

theory advocated for structures, could be applied to any material at all, it would be more applicable to concrete than to steel.

178. Nevertheless, a mere titular alteration could hardly change the content of a theory. On the other hand, the necessity for such a change was significant in itself, because this indicated the fluidity of the foundation underlying the theory.

179. Under the heading "Design criteria for frame", no design criterion could be found. What were mentioned were failure criteria (Causes of failure), which appeared to have been identified with design criteria.

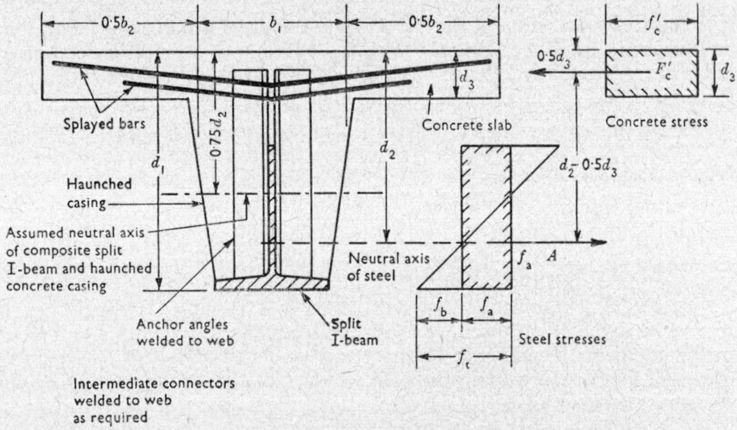

FIG. 29B.—COMPOSITE TEE-BEAM

180. Whenever an engineering design was made, the designed object was expected to work successfully or to satisfy certain requirements, which were usually called the design criteria. Failure was an anomaly to be avoided, not an aim to be achieved; therefore it could only form a basis of checking and could not replace the design proper.[37]

181. Such philosophical considerations might at first appear pedantic, but as the following discussion would show, they led to a methodological impasse and made it necessary to incorporate contradictory assumptions in the "theory".

182. Probably it was not far from true that few reinforced concrete structures under load behaved in the elastic manner normally assumed. However, was there any concrete structure in reality behaving in the plastic manner "normally" assumed?

183. It did not seem to be impartial to compare strain measurements on actual structures made three decades ago with load measurements made today. First, conditions in laboratory tests, from which ultimate loads were derived, could hardly reproduce the real behaviour of a structure; therefore the laboratory accuracy of ultimate load did not mean much.

184. Secondly, factors influencing strain measurements were much more delicate and complicated than those influencing load measurements. For instance, strain implied, and was handicapped by, the assumption of stress distribution and homogeneity.

185. Thirdly, the departure from the elastic manner normally assumed had apparently been wrongly diagnosed, because recent findings[44] had refuted this premature conclusion, which was often used to discredit elastic methods.

186. Finally, if some strain or load measurements were made on a plastic-designed structure at working load, should the result bear a close relation to values suggested by any ultimate load theory? If so, the structure was reaching its ultimate state and its designer his ultimate fate. If not, was it fair to amplify the lack of agreement in elastic

design while no agreement of any sort had ever been conceived in the ultimate load theory?

187. The ultimate load was alleged to be essential for determining the true margin of safety available against failure. Meanwhile, the ultimate and the working load would be so determined that their distribution, and rate, frequency and duration of application had the worst characteristics which could occur in relation to failure. These statements were rather confusing, because such highly technical terms had been used in their most primitive sense.

188. On the other hand, the most important characteristic of loading, namely its intensity, was not specified at all. Was the load mentioned under the heading Definitions the mean, the mode, the medium, or a novelty in terminology? (For further discussion, see Ref. 45.)

189. The effects of such terminological laxity immediately manifested themselves in the well known load factor, which was defined in the Report as the factor by which the appropriate working load was multiplied to obtain the design ultimate load. It was difficult to interpret a noun which was defined only in terms of vague, undefined, and often meaningless other nouns.

190. The state representing the ultimate load was no more than sheer wishful thinking. After being rescued from the point of collapse by a load factor, the ultimate-load-designed structure would never, at least its owner hoped, meet its ultimate fate. Therefore, such a structure under normal loading would never fail and thus repudiate the "mechanism" on its designer's drawing board. On the other hand, this structure might fail under a set of abnormal loading. In short, an ultimate-load-designed structure could only behave either normally abnormal or abnormally normal.

191. The method adopted in the Report for determination of the load factor was obsolete, and the statistical manipulation involved in it was artistic.

192. It was rather revealing that an articulated system of structural members connected by hinges was defined as a "mechanism" without reference to its topology. If this definition were accepted, then the structure analysed and designed by the advocated method was not a structure after all, but a mechanism capable of rather dynamic performance. Furthermore, to restrict the magnitude of deformation according to whether some equations were materially affected seemed to be highly academic.

193. Although the design method had changed its name from "plastic" to "ultimate load", its true nature was exposed again whenever the term plastic hinge made its appearance. The fortune of plastic theory was indeed hinged upon the concept of the plastic hinge.

194. The legend of Fig. 5a indicated that the depicted hinges, which were later identified with plastic ones, were "frictionless hinges for statical determinacy". This had several interesting implications.

195. First, a plastic hinge was frictional by nature, and could no longer be plastic if it became frictionless. This was clearly exhibited by the so called St Venant's model, but had been habitually overlooked in the plastic theory. Among all methodological errors committed by plasticians, this was perhaps the most fatal, and had led to the establishment of inconsistent, hence fallacious, theorems claimed to be the foundation of the plastic deformation theory.[46, 51]

196. Secondly, the "frictionless hinges" seemed to be required only "for statical determinacy". Therefore, plastic hinges and their associated mechanism were professed as an imaginary artifice in analysis instead of a physical phenomenon in reality. This struggle for statical determinacy could be traced back to the birth of Van den Broek's limit design, which was formulated long before the age of machine computation, and symbolized the defeatism of structural analysts.

197. Thirdly, because of the astronomical number of possible incomplete, complete, and overcomplete mechanisms (See Ref. 52 for example) a rather artistic recommendation on the choice of positions of hinges became necessary. From the "determinacy

struggle", it was clear that the ultimate load method for designing a statically indeterminate structure was in fact a method for designing a statically determinate structure with arbitrarily fixed frictional hinges. The positions of such hinges were either recommended or arbitrarily picked out from a billion possibilities by the designer.

198. This scheme would work, and a similar situation in arch design sometimes occurred: a designer only acquainted with three-hinged arches would insert "hinges" even to a fixed arch for statical determinacy. The resulting arch would stand up, but would be heavier than necessary. Incidentally, the positions assumed for hinges shown in Fig. 14 were rather misleading. The three hinges situated so near to the junction would behave as one.

199. In almost every detail at the foundation level, the yield-line method for slabs was similar to the ultimate load method for flexural frames. A concrete slab was merely decomposed, and thus became a mechanism with frictional line-hinges.

200. The only fundamental difference in these two theories was that the mechanism in the yield-line theory was real whereas that in the ultimate load theory was but a statically determinate pseudo-mechanism. If a real frictional mechanism could be accepted for simulating slabs, why could it not be used for simulating frames?

201. At the philosophical level, the yield-line theory suffered as much as the ultimate load theory for frames from the same incurable ailment. For a real slab, to yield meant that the slab had been badly designed; and not to yield meant that the yield-line design bore no relation to its counterpart in reality. Both prospects seemed to be equally gloomy.

202. In § 121, the moment transfer allowed in design codes had been accredited to the ultimate load method. Although this might be the original intention, the actual effect had been quite different.

203. According to the plastic theory, redistribution of bending moments and transfer of support moments, etc. could only occur when sufficient plastic hinges had formed and the mechanism had been mobilized. Should a structure designed by the plastician really behave like a mechanism? If not, how could such redistribution and transfer of moments be effected?

204. The reason why structures designed in this way were still standing and had not yet become mechanisms (a discredit to the plastic design!) was that apart from the convenient escape through the load factor, a structure would behave as the designer wanted it to, either intentionally or unintentionally.

205. The effect of transferring a certain amount of moment was the same as that of using a different loading system, or a different responding system, or both to produce such a "transferred" moment pattern. Since the original loading was quite arbitrary, there was no reason why a new loading system could not be used.

206. According to the Report, tests to prove the validity of the applied yield-line theory for slabs had been carried out by many investigators. However, the validity of such tests had not been proved, if it could be proved at all.

207. It was insufficient to obtain test results. What was far more important was their interpretation, amplification, and application. Even when the reported good agreement between the test results and the calculated ones was accepted, this confirmed no more than a verification of some perfection of test technique.

208. The validity of tests could be proved only if the conditions existing in reality could be faithfully reproduced in tests. Unfortunately, the actual structure and the real loading were much more complicated than their counterparts in tests. The crucial point was that there was, or at least should be, no plastic hinges, mechanism and other fantasies in the actual structure, whereas such things invariably appeared in plastician's tests.

209. Mr Su felt that his discussion was by no means as comprehensive as would be desirable. Nevertheless, the history of science had shown more than once that only a few "minor" inconsistencies in a theory were sufficient to spark off an explosion. If there existed so many major fallacies in it, the theory and its associated methods could

hardly be regarded to be of positive value. It only provided a hard lesson that experiments and mathematics were merely servants of the engineer and should not be allowed to usurp the power of a master.

Mr R. E. Rowe (Head, Design Research Department, Cement and Concrete Association) commented that in the introduction it was stated that the purpose of the Report was to attempt to clear the way for a logical treatment of the structure as a whole. This was certainly necessary when considering design procedures at the present time but it appeared to Mr Rowe that while the committee had clearly stated an approach with regard to the initial stages of this process, it had not emphasized sufficiently the importance of the subsequent stages, nor had the necessary research been defined, which would enable these stages to be fulfilled.

211. Considering the idealized moment/curvature relation for beam or column sections that suggested in Fig. 1 was perfectly reasonable and the parameters used to define it might, if necessary, be modified in the light of the research at present being carried out all over the world and being co-ordinated by the European Committee for Concrete. However, to apply it to a framed skeletal structure in the manner outlined in the Report was only satisfactory provided that sufficient experimental evidence existed, not so much for the moment/curvature relation for a section, but the same relation for the various types of joint which occur between members in the frame. Very little experimental data existed on this subject or, for that matter, on tests on frames in general. The need for such data was fairly obvious and the committee could well have stressed this aspect.

212. However, the skeletal structure was by no means the structure as a whole and in any framed structure the interaction between the frame (which had been discussed in detail in the Report as a separate entity), the floors and the shear walls or cladding was also of vital importance. Considering the structure as a whole, the interaction effects of the various parts of the structure on the moment/curvature relation at the critical sections and also on the position of the necessary plastic hinges might be considerable and, if considered, could possibly lead to a simplification in the limit analysis. Interaction between bending and twisting moments would also exist and this interaction had a very significant effect on the rotation capacity in torsion of prestressed concrete and probably of reinforced concrete sections. These aspects should obviously be considered if a logical approach to design was to be produced and, as yet, insufficient data was available to allow any precise recommendations to be made.

213. Mr Rowe thought it necessary to emphasize the urgent need for research in this field so that the ultimate load design approach to concrete structures as a whole would then become, as it must do, generally accepted. In his opinion the interaction effects would probably enable simplified limit analysis techniques to be adopted, even for complicated structures, with advantage to the designer. If, therefore, a relatively complicated design approach, which was not rational for the entire structure, although a little more logical than the existing simple design approach, was introduced at this stage, it would not be accepted by designers and might well build up a resistance to limit design. This would retard progress in structural concrete design, which was contrary to the object of the committee.

214. With regard to the more specific points in the Report, in § 51 the values of k_1 and k_3, used in defining the plastic length 1_p, were based upon a limited number of tests. It seemed necessary for more experimental results to be obtained before precise values of k_1 and k_3 were given, particularly for the hot-rolled high tensile deformed bars now becoming available, and for conditions of sustained load. It was also stated in the Report that in practice 1_p would lie in a range from 0·4 to 2·4*d*. This range would be considerably affected by the applied loading and moment between points of contraflexure and, for simply supported beams with single point loads, values of 1_p of 2·4 had been found with cold-worked reinforcing steel. Hence, in practice, it seemed likely that the upper limit to the value of 1_p could well be considerably greater than that given.

215. Dealing with shear, § 91 stated that the formation of a plastic hinge might reduce the shear resistance of a reinforced concrete section. While this was undoubtedly true, surely it was of more interest to know how the presence of shear decreased the rotational capacity of the section? It was on this aspect that the research was required.

216. In §§ 95–120 dealing with the application of yield-line theory to slabs, the slabs had been considered as a separate unit and the treatment given was extremely brief and unenlightening from the practical designer's point of view. If the amount of information to be given was limited, it might have been preferable to give the basic properties of yield lines, the two fundamental approaches of virtual work and equilibrium, giving as they did upper- and lower-bounds solutions, and to have mentioned the possible effect of membrane action that can occur in slabs. In yield-line theory the use of affinity relations in transforming slabs was extremely difficult to follow in the existing literature; guidance on this method, when dealing with orthotropic reinforcement, which was the only means of utilizing the equations given in the Report, would have been welcome.

217. Mr Rowe was interested to read that shake-down could occur in slabs and would like to know more of the evidence that supported this contention. The whole problem of slabs and columns was of extreme interest in view of the frequency with which flat plate structures were now used, and such problems as the effect of punching shear, curtailment of the reinforcement on the yield-line patterns, and methods of calculation, and the optimum distribution of reinforcement needed investigation. It was rather surprising that little attention had been paid to this aspect of ultimate load in the Report.

218. Finally, since it was vital to the ultimate load design procedure, Mr Rowe thought that more attention might well have been given to the selection of the appropriate load factors. Methods of analysis under ultimate load conditions were continually improving; so was knowledge of the fundamental properties of the materials used in reinforced concrete and prestressed concrete, but very little scientific information was available about the magnitude of, and the variation in, the live loads applicable to the different types of structure which had to be designed. In this field there was considerable scope for research to provide one of the principal criteria used in defining the load factor and possibly with the formation of the Civil Engineering Research Council, there now existed a means whereby this type of research could be encouraged.

Mr L. L. Jones (Research Engineer; Cement and Concrete Association) wrote that if a report on ultimate load analysis were to be reasonably complete it should cover the following three aspects:

(*a*) the assessment of the load factor;

(*b*) the analysis of the structure, whereby the forces and moments at failure were assessed;

(*c*) the strength-of-sections equations which were used to proportion the individual members to fail under the force and moment system calculated in (*b*).

220. While some of the Report was devoted to items (*a*) and (*c*), the majority of it was concerned with one aspect of item (*b*), namely, the analysis of skeletal frames. It could not be denied that this was an important field of ultimate load analysis, nevertheless the overall impression of the Report was that other equally important aspects of the subject had not received the same degree of attention.

221. In particular the Report did not contain a single equation which could be used to assess the strength of individual members subjected to a combination of force and moment, and this was surely a serious omission? It was of course true that equations for members subjected to bending, or bending and axial load were already contained within British Codes of Practice; however, further amplification and comparison with current practice abroad would have been of interest. British codes did not, however, contain equations for members subjected to bending and shear or bending and torsion,

and unfortunately the committee had not offered any guidance in this field. These two subjects, of course, were not yet fully understood but even safe interim equations would be of great assistance to the design engineer.

222. The main part of the Report presented a method of analysis for skeletal frames, which was clearly a genuine attempt to ensure that members would have sufficient rotation for the structure to become a mechanism. However, since this method was put forward and the method usually associated with steel frames had been disregarded, perhaps the committee could have given their reasons for this decision. Certainly Mr Jones knew of no large series of tests which proved that frames failed at a substantially lower value of load than that calculated by the Cambridge method.

223. With regard to the assessment of a load factor the committee presented two alternative methods of approach which included what were generally accepted to be the items which should influence the magnitude of this factor. The introduction of either of these two methods into British Codes of Practice would, however, cause certain human problems. If any system was to prove successful it must be in such a form as to eliminate as far as possible in an assessment, decisions which were purely a matter of opinion. This was particularly important since calculations were frequently checked by a local authority which would find itself placed in the unenviable position of possibly disagreeing with an engineer's assessment of a load factor purely on the basis of opinion.

224. If the recommendations of the European Committee were considered, it could be seen that the total load factor was governed by three main items, the "overload" factor, the assumed strength of the materials and the seriousness of collapse. At present the suggested method of assessing the overload factor was unworkable since the statistical data required was not available. Instead therefore, the committee might consider tabulating the working load and overload factor to be taken for a certain class of structure in a similar way to that reported by Yu, Corbin and Hognestad.[54] For the strength of the materials a standard value could be adopted for the steel stress and for the concrete stress it was only necessary to ensure that the value assumed in the calculations was achieved later on site. With regard to the seriousness of collapse again, structures could be classed into groups and the appropriate multiplying factor tabulated. If presented in this form the same basic method of assessment could be employed but the decisions concerning the magnitude of the factor would have been tabulated by a committee. This would have the advantage of maintaining the rational basis of the variable load factor while ensuring that a certain minimum load factor would always be achieved.

225. Since statistical surveys of loads on buildings were referred to in the Report, perhaps the committee would consider making the strongest possible recommendations for research into loading intensity to be undertaken on a national scale.

226. The section dealing with the analysis of slabs was felt by Mr Jones to be rather curtailed and while the information was correct, many aspects could have been amplified. With regard to the equations given for nodal forces, it should be noted that orthotropic slabs need not necessarily be transformed into isotropic slabs before they could be solved. Equations for nodal forces in orthotropic slabs[55] could be used, thus avoiding the theorems of affinity.

227. When a designer had analysed an orthotropic or continuous slab he must choose the ratio of the moments in the various directions and on this choice would depend to a great extent the amount of redistribution necessary. This would in turn affect the serviceability of the slab. It seemed common sense to try to have the ratios the same as the elastic moments ratios but little experimental evidence seemed to be available to show how far one could depart from this. Had the committee any information on this point from current practice overseas?

228. Another important aspect of slab design was the extent of top reinforcement in continuous slabs. Although equations could be formulated from which this distance could be calculated, as far as Mr Jones knew, these equations had never been substantiated by tests. If possible, therefore, research into this field should be encouraged.

Mr M. Z. Cohn (Research Engineer, Building Research Institute, Bucarest) wrote that his contribution was concerned only with some comments of the main items and with an alternative approach for the design of continuous beams.

Loading schemes

230. The Report stated in § 26 that the application of the total load for a short time was a reasonable criterion for ultimate conditions. It was specified, however, that "the designer must judge whether the nature of the structure and the live/dead load ratio might be such that some form of partial loading provided a worse condition".

231. If only ultimate load considerations were to be allowed for in structural design, full loading would be fully accurate, while resulting in large simplifications of the numerical calculations.

232. Unfortunately reinforced concrete structures could not be designed for ultimate conditions exclusively (a fact indicated by § 72 of the report) and partial loading could occur during their working life.

233. Consideration only of the full load in the ultimate load design of concrete structures might lead to defective working conditions, consisting in unequal yield safety of critical sections with respect to the corresponding worst loading arrangement, and eventually in premature formation of plastic hinges (the limit L_1 was reached at working loads).

234. Therefore partial loading did not appear as an occasional case, consideration of which might be left to the designer's judgement, but was a necessary case to be taken into account in order to ensure a satisfactory structural behaviour in the working range. This remark became the more important, the smaller was the dead/live load ratio.

Load factors

235. Reflecting modern trends, the formula for L.F. in § 66 gave in principle a sound basis to the solution of the problem. In Mr Cohn's opinion such an assessment of the L.F. was a sensible progress in allowing for the variability of parameters affecting the ultimate safety of a structure. Obviously the weights suggested could be the subject of further discussion and refinement.

236. Throughout the Report (the examples included) the L.F. multiplies the total (dead+live) load. In view of the progressive background involved in the assessment of L.F., this procedure appeared to be unrealistic, though conservative.

237. In fact the L.F. must reflect in some logical way the probability of overloading. It would be contrary to this tendency to provide the same probability of overloading for the dead and live loads, i.e. both to be multiplied with the same L.F.

238. In recent design provisions distinct values were recommended for the dead and live loads. Thus in the Soviet norms[56] values of 1·1 and 1·4 were suggested, variability of materials being considered independently through "homogeneity factors" affecting the standard stress values of steel and concrete. Currently these factors were taken as 0·9 and 0·55 respectively.

239. In the proposed revision of the American Concrete Institute Building Code 318–66[57] L.F. values of 1·3 and 1·56 were given for dead and live loads, the homogeneity factors implied being 0·8 and 0·67, for steel and concrete respectively.

240. As shown in § 63, even BSCP 115:1959 followed this trend recommending L.F. values of 1·5 and 2·5 for dead and live loads respectively, and a L.F. of 2 for overall loading. This point of view was adopted in other studies.[58, 59]

241. In conclusion, recommendation of a distinct L.F. according to the type of loading appeared to be justified. A simple and logical assumption would be to consider that the intensity of dead loads could not vary and was exactly known (even if an allowance of 5 to 10 per cent were included for the accidental increases in the specific weight of materials or in the dimensions of members versus those designed). The L.F. would thus affect only the live loads. For very small or no live loads a reduced overall L.F. could be provided.

Approaches to ultimate load design

242. The ultimate load design for concrete structures implied three basic requirements: (1) limit equilibrium, (2) rotation compatibility, and (3) serviceability.

243. Irrespective of the criteria to meet these requirements, usually only two of the three conditions could be satisfied initially, the third being the object of subsequent check.

244. Two distinct approaches were possible in this connexion: a first approach, in which the design was based on limit equilibrium and compatibility, serviceability conditions being given a separate consideration; a second approach in which limit equilibrium and serviceability were accounted for from the very initial steps of the design, while compatibility required a special check.

245. The first approach, promoted by Professor A. L. L. Baker and his co-workers, was the basis both for the Report under discussion and for the work of Committee XI of the European Committee on Concrete (CEB).[60, 61]

246. The second approach had recently been used by Petcu[62] and Cohn[63] and seemed to yield good results, part of which would be described later.

247. The following aspects were to be emphasized with regard to the approach used in the Report. It was essentially one of analysis, since both the concrete dimensions and the amount of steel reinforcement must be ascertained in order to know the $\bar{X}_k$ values and to see whether or not the compatibility equations were verified. Generally a trial and error procedure was needed to obtain ultimate rotations of plastic hinges less than the permissible rotations, as required.

248. In evaluating the various methods proposed in the Report, idealizations accepted in the simplified method (§ 37) appeared as reasonable, safe approximations for the actual behaviour of reinforced concrete structures.

249. The rigorousness claimed for the methods indicated in §§ 28 and 29 would be a reality only if material constants, on which they were based, could be known with the corresponding degree of accuracy, as might happen in research work.

250. In current design, however, the possibility of assessing correct M/ψ curves and to know the actual EI values at various loading stages of all plastic hinges and members was highly uncertain, if not a fiction.

251. Mr Cohn thought therefore that the accuracy gained from "the rigorous methods" was more theoretical than real. A similar situation was probably to be expected for any procedure based on parameters of such a large scatter of test results as the E_cI values are.

Shear provisions

252. In § 90 of the Report, the recommendation of CP 115:1959 is mentioned, according to which all major structural members should have shear reinforcement.

253. Adoption of this recommendation would be welcomed in the ultimate load design of concrete structures. For the time being perhaps even more conservative measures would be useful to avoid any risk of fracture in diagonal tension. In many cases certain savings of shear reinforcement could be obtained as a result of the moment redistribution, leading to ultimate shears less than the elastic shears calculated for the factored loads.

254. However, in view of the present status of knowledge, it would be simpler and safer to neglect the shear redistribution. By so doing the basic requirement of the ultimate load design (failure to occur by bending and not by shear) is implicitly fulfilled.

255. At present therefore it seemed advisable that structures designed for the ultimate load with respect to bending be provided with transverse reinforcement (irrespective of the working shear stress) based on the elastic shears due to the factored loads.

Design of continuous beams

256. Some peculiarities of the rule for the design of uniformly loaded continuous beams (§ 137) were outlined below. The rule is too limited, since it applies only to

uniform loading, to simply supported ends, and probably to beams with more than four spans. No procedure is given for other usual loadings, for the case of end fixity of the beam, and for a smaller number of spans.

257. Even with these limitations the rule should give more clear indications for the moment value at the first intermediate support, the choice of which is left to the designer. For the remaining critical sections, equalization of moments at the inner spans and supports is recommended, but it would be more specific to suggest the moment value as $K(G+P)l/16$.*

258. It would be desirable that the condition $\theta_p > 0{\cdot}16Ml/EI$ be put in a simpler form for design purposes, i.e. such that no additional computations be needed besides those involved in proportioning the steel reinforcement. For instance Soviet specifications require the reinforcement of critical sections to be such that $n_u > 0{\cdot}3$, as indicated in § 129.

259. The proposed rule implies application of the total dead and live $(G+P)$ loads and of an overall L.F. affecting it.

260. As a corollary, if partial loading is possible in practice, the given rule leads to variable yield safety for the critical sections and, under certain circumstances, even to formation of plastic hinges at working loads.

261. As an illustration, consider a five-span continuous beam designed according to the rule in § 137. Support sections are denoted as 0, 2, 4 and span critical sections as 1, 3, 5. Let G and P be the total dead and live loads for each span respectively, K the adopted ultimate L.F. and K_1 the yield load factor (i.e. that under which the first plastic hinge occurs at the critical section considered). Under the ultimate load $K(G+P)$ the central span collapses while the plastic moments at sections 4 and 5 are

$$\bar{X}_i = \bar{X}_4 = \bar{X}_5 = \alpha K(G+P)l \quad . \quad . \quad . \quad . \quad . \quad . \quad . \quad (13)$$

where according to § 137 $\alpha = 1/16$.

262. The loads being increased proportionally, assume that various arrangements of the live load are possible. In the elastic range (OL_1) the moments will be:

$$M_i = a_i Gl + b_i Pl \quad . \quad . \quad . \quad . \quad . \quad . \quad . \quad . \quad . \quad (14)$$

a_i and b_i defining the maximum elastic moment at every section 1. When this section becomes the first plastic hinge (point L_1) the plastic moments are:

$$\bar{X}_i = K_{1i}(a_i Gl + b_i Pl) \quad . \quad . \quad . \quad . \quad . \quad . \quad . \quad . \quad (15)$$

263. From (13) and (15) the ratio of the yield to the ultimate safety (the yield safety parameter) is:

$$x_i = \frac{K_{1i}}{K} = \frac{\alpha(G/P+1)}{a_i G/P + b_i} \quad . \quad . \quad . \quad . \quad . \quad . \quad . \quad . \quad (16)$$

264. A condition not to have plastic hinges at working loads is $K_1 > 1$, or from relation (16) $1/x_i < K$. But for proportional loading $1/x_i = r$ is the redistribution factor of the beam, i.e. the ratio of ultimate to first yield loads. Hence the condition that plastic hinges do not form in the working range can be expressed as

$$r < K \quad . \quad . \quad . \quad . \quad . \quad . \quad . \quad . \quad . \quad . \quad . \quad (17)$$

a result which has been stated in an earlier paper by the writer.[64]

265. Making now $a_4 = 0{\cdot}079$, $b_4 = 0{\cdot}111$, $a_5 = 0{\cdot}046$ and $b_5 = 0{\cdot}086$ in relation (16) values of x_4 and x_5 can be derived and plotted against G/P as shown in Fig. 30.

266. On the curve of x_4 $r = 1/x_4 = K$ values are indicated for $K_1 = 1$. Therefrom it is seen that for $G/P \simeq 0$ and K less than 1·78 a plastic hinge appears at section 4 at service loads. For $G/P = 0{\cdot}5$ this happens when L.F.$=K$ is less than 1·6.

267. Care must therefore be given to the behaviour of intermediate supports for small values of G/P and for L.F. less than 2, when assessed acording to § 66.

268. Fig. 30 also shows that x_5 being larger than x_4 service conditions are better

* The additional notation is listed at the end of Mr Cohn's discussion on pp. 66 and 67.

for the interior spans than for the interior supports, and this is quite normal. However for G/P larger than about 1·5 it is seen that x_5 exceeds unity, that is the design according to § 137 yields larger values of the span moments than the elastic design by using relation (15).

269. Finally, inspection of eq. (16) shows that a design for similar working condition (equal x_i values) for all span and all support sections respectively is not possible, except for certain particular G/P values.

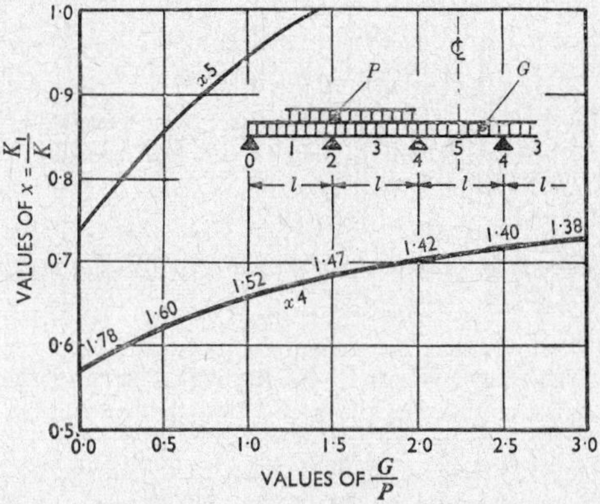

FIG. 30

OPTIMUM LIMIT DESIGN OF REINFORCED CONCRETE STRUCTURES

Assumptions and basic considerations

270. The approach which follows will be shortly referred to as an "optimum limit design". Limit design implies the assumptions (1) and (2) below, while "optimum design" implies the simultaneous fulfilment of certain rational limit and serviceability criteria. Evidently further optimum solutions can be imagined, according to the criteria adopted.

271. In this connexion two optimum limit design procedures will be presented. In the first procedure the ultimate load factors are constant for all elementary mechanisms, the yield load being generally unequal for various critical sections. In the second procedure L.F. are variable and yield load factors are assumed to be constant.

272. Contrary to the theory involved in the Report both "optimum procedures" are not concerned with the compatibility problem, which will be dealt with separately.

273. The main assumptions involved in the "optimum limit design" are as follows:

(1) The behaviour of reinforced concrete members can be idealized by the bi-linear diagram $OL_1\ L_2$ adopted in the Report.

(2) Failure occurs when at least one span of the beam is turned into a mechanism.

(3) Members resist in bending, convenient measures preventing local excessive strains in concrete, whether from shear failure or buckling.

(4) Live loads may have any possible location, so that the worst combination can be considered for each critical section in turn.

(5) The load factor affects only the live load, while the dead load (including a 5 to 10% allowance for casual increases) is assumed to remain unaltered during the loading process.

274. Contrary to the methods suggested in the Report the plastic moments $\bar{X}_i$ at all the critical sections are *not* known and to find their values so that limit and serviceability requirements be fulfilled is the object of the "optimum design" approach.

275. The general form of the $\bar{X}_i$ value in accordance with the above assumptions can be derived from the preliminary remarks which follow.

276. Consider the moments in one of the critical sections i at three successive stages of loading, for the particular arrangement causing the first plastic hinge to occur at i:

(*a*) in the working range (between points O and L_1) the elastic moment being given by relation (14)

(*b*) at yield (point L_1) the live load being increased to $K_{1i}P$

$$\bar{X}_i = a_i Gl + K_{1i} b_i Pl \quad (18)$$

(*c*) at collapse (point L_2) the live load reaching the ultimate value KP

$$\bar{X}_i = a_i Gl + K\bar{b}_i Pl \quad (19)$$

277. According to the idealized behaviour accepted, the plastic moment $\bar{X}_i$ is invariable from L_1 to L_2, while the load increases from $K_{1i}P$ to KP. The unknown parameter $\bar{b}_i$ results from (18) and (19) as:

$$\bar{b}_i = b_i K_{1i}/K = b_i x_i \quad (20)$$

so that the plastic moment at section i can be expressed in the final form:

$$\bar{X}_i = a_i Gl + b_i \cdot x_i \cdot KPl \quad (21)$$

278. Relation (21) shows that the design for ultimate loads implies that only the yield safety parameters $x_i = K_{1i}/K$ are to be known, since the remaining factors in relation (21) are the same as those of the elastic design.

279. Now s being the number of critical sections ($i = 1, 2 \ldots \ldots s$) and m the available number of elementary mechanisms (excluding of course the fictitious joint mechanisms) m limit equilibrium equations may be expressed in terms of a convenient *limit criterion*. Since generally $s > m$, $s - m$ additional relations are needed in order to complete a set of s equations and to determine all the s unknown values of x_i. These s-m relations are provided by an appropriate *serviceability criterion*.

First optimum procedure[63]

280. Assume as limit criterion that all possible mechanisms have the same ultimate safety. Then it can be shown that the limit equilibrium equations may be written in the general form:

$$\sum_i \lambda_{ik} x_i \bar{M}_i = \bar{M}_{ok} \quad (22.1)$$

the sum being extended to all the plastic hinges of the k-th mechanism. If all $\bar{M}_i$ and $\bar{M}_{ok}$ are expressed in terms of single loading, and geometrical parameters relation (22.1) simplifies to

$$\sum_i \lambda_{ik} x_i b_i = \phi_k \quad (22.2)$$

281. It is to be noted that the limit equilibrium equations (22.1) and (22.2) involve $\bar{M}_i$ and b_i values referring *only* to the elastic moments due to the live loads.

282. As serviceability criterion the equality of $s - m + 1$ values of x_i will be accepted, so that the remaining $m - 1$ ones be larger than the common value $\bar{x}_o$.

283. Indexing the critical sections in a convenient order the serviceability criterion is expressed by the following relations:

$$x_i = \bar{x}_o = \text{const. (for } i = 1, 2 \ldots\ldots s - m + 1) \quad (23.1)$$

$$\text{and } x_i > \bar{x}_o \quad \text{(for } i = n + 2 \ldots\ldots s) \quad (23.2)$$

284. It has been proved[63] that $\bar{x}_o$ can be derived by minimizing the balanced unknowns $\bar{x}_k$ in each of the eqs. (22), i.e.:

$$\bar{x}_o = \min \bar{x}_k = \min \left[\frac{\phi_k}{\sum_i \lambda_{ik} \bar{b}_i} \right] \quad (24)$$

285. Substituting the value of $\bar{x}_o$ in the remaining $m-1$ eqs. (22) a single unknown is to be determined from each equation, the problem being thus solved.

286. The above criteria and the procedure described are particularly suited to structures failing in beam-mechanisms. Design tables are already in existence for continuous beams.[63, 65]

287. Extension of the same principles to frames is possible, but some refinements must previously be made to allow for the effect of axial forces. When this effect is negligible, application of the technique outlined to frames raises no difficulty.[63]

288. As a further illustration example 3 in the Report will be redesigned by the procedure presented. For simplicity it will be accepted that the load factor applies to the total load, as in the Report. (Then the optimum design applies as the dead load would be negligible versus the live load.) However all the possible partial loadings will be considered.

289. The maximum (minimum) elastic moments are first derived. For $I_b h/I_c l = (27/18)^3.15/25 \simeq 2$ the moment constants in the five fundamental loading cases are found from usual handbooks,[66] as shown in Fig. 31.

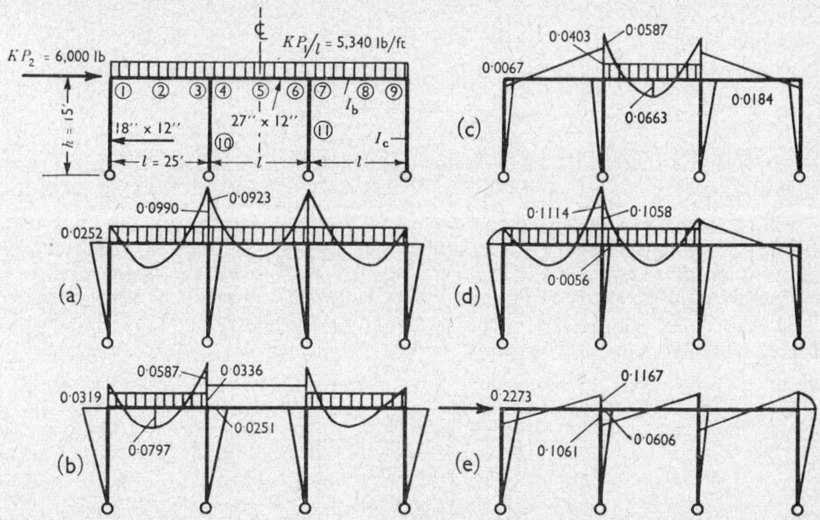

FIG. 31

290. For the cases (*a*), (*b*), (*c*), (*d*) the elastic moments for the factored load are $\bar{M}_i = b_i KP_1 l = 40 \times 10^6 b_i$ lb-in, while for case (*e*) the moments are $\bar{M}_i = b_i KP_2 h = 1{\cdot}08 \times 10^6 b_i$ lb-in.

291. From the loading combinations indicated in parentheses the moments (reduced by 10^6 lb-in) were found to be:

$$\bar{M}_1 = 1{\cdot}27\ (b);\quad \bar{M}_2 = 3{\cdot}25\ (b+e);\quad \bar{M}_3 = 4{\cdot}64\ (d+e);\quad \bar{M}_4 = 4{\cdot}23\ (d);$$
$$\bar{M}_5 = 2{\cdot}65\ (c);\quad \bar{M}_6 = 4{\cdot}35\ (d+e);\quad \bar{M}_7 = 4{\cdot}45\ (d);\quad \bar{M}_8 = 3{\cdot}19\ (b);$$
$$\bar{M}_9 = 1{\cdot}52\ (b+e);\quad \bar{M}_{10} = 1{\cdot}07\ (b+e);\quad \bar{M}_{11} = 0{\cdot}80\ (c+e)$$

292. These are the envelope moments used in the elastic design if sections are proportioned according to the ultimate strength theory and are illustrated in Fig. 32.

293. Four limit equilibrium equations can be written, corresponding to the three

beam-mechanisms and to the panel-mechanism available. With $\lambda=0{\cdot}5$ (since span plastic hinges are assumed at mid-spans) eqs (22.1) are:

$$0{\cdot}5\times1{\cdot}27x_1+3{\cdot}25x_2+0{\cdot}5\times4{\cdot}64x_3=5 \quad (25.1)$$

$$0{\cdot}5\times4{\cdot}23x_4+2{\cdot}65x_5+0{\cdot}5\times4{\cdot}35x_6=5 \quad (25.2)$$

$$0{\cdot}5\times4{\cdot}45x_7+3{\cdot}19x_8+0{\cdot}5\times1{\cdot}53x_9=5 \quad (25.3)$$

$$-1{\cdot}27x_1+1{\cdot}52x_9+1{\cdot}07x_{10}+0{\cdot}80x_{11}=1{\cdot}08 \quad (25.4)$$

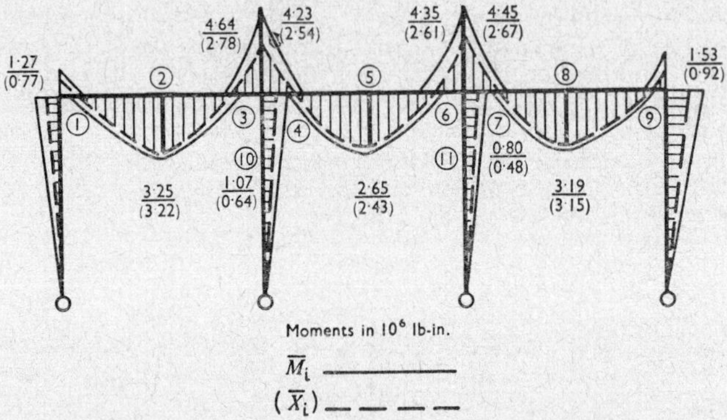

FIG. 32

294. Balancing in turn the unknowns in each equation results in $\bar{x}_{\mathrm{I}}=0{\cdot}81$, $\bar{x}_{\mathrm{II}}=0{\cdot}72$, $\bar{x}_{\mathrm{III}}=0{\cdot}81$ and $\bar{x}_{\mathrm{IV}}=0{\cdot}51$. The minimum value is $\bar{x}_0=0{\cdot}51$ from eq. (25.4). Since substitution of $x_1=x_3=\bar{x}_0=0{\cdot}51$ in eq. (25.1) yields $x_2=1{\cdot}07>1$, the larger value $\bar{x}_0=0{\cdot}6$ is selected (Fig. 33), resulting in $x_2=0{\cdot}99$, $x_5=0{\cdot}92$ and $x_8=0{\cdot}99$ from eqs (25.1), (25.2) and (25.3) respectively. The final plastic moments $\overline{X}_i=x_i\overline{M}_i$ are specified in Fig. 32 by dotted lines.

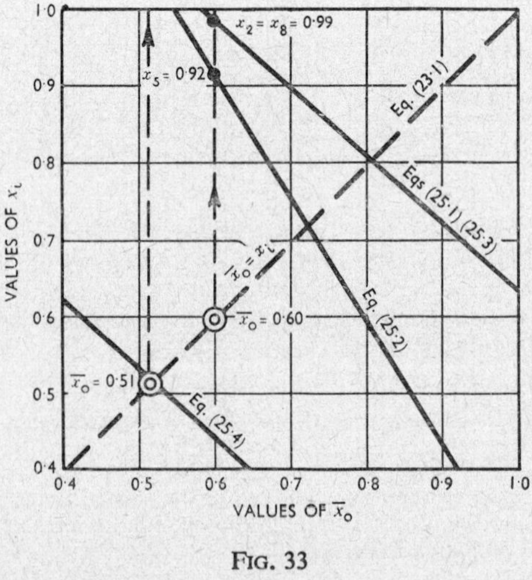

FIG. 33

295. A geometrical interpretation can be given to this solution. Making $\bar{x}_0=x_1=x_3=x_4=x_6=x_7=x_9=x_{11}$ each eq. (25) may be represented as a line in the plane $(\bar{x}_0, x_i)$. Fig. 33 shows the lines corresponding to these equations. Inequality (23.2) shows that the intersection point of the bisectrix (23.1) with the lines (25) yielding the minimum $\bar{x}_0$ is the required solution. On the diagram in Fig. 33 $\bar{x}_0=0{\cdot}51$ is found, but it is seen that $x_1>1$ follows. For $\bar{x}_0=0{\cdot}60$, however, $x_2=x_8=0{\cdot}99$ and $x_5=0{\cdot}92$ are found on lines (25.1) and (25.2).

Second optimum procedure[67]

296. In the above procedure the primary emphasis was placed on the limit criterion, constant ultimate load safety being assessed with variable yield safety resulting.

297. A modified way can be followed placing the main emphasis on the serviceability criterion, providing equal yield safety to desired critical sections, and varying the ultimate safety of the mechanisms.

298. The following basic criteria are adopted:

Limit criterion: a minimum ultimate safety is ensured for each mechanism, but not necessarily the same for all mechanisms.

Serviceability criterion: equal yield safety for all supports, equal yield safety for all spans, but larger yield safety for span than for support critical sections. No plastic hinges at working loads.

299. For continuous beams (the support critical sections being 0, 2, 4 . . . . 2j and the span sections 1, 3, 5 . . . $2j+1$) it is required:

$$x_0 = x_2 = x_4 = \ldots\ldots = x_{2j} \qquad (26.1)$$

$$x_1 = x_3 = x_5 = \ldots\ldots = x_{2j+1} \qquad (26.2)$$

$$x_0 < x_1 \qquad (26.3)$$

$$1/K < x_0 \qquad (26.4)$$

300. The general equations (22.2) turn into inequalities with only three terms, the form of which is

$$(1-\lambda)b_0x_0+b_1x_1+\lambda b_2x_2 \geqslant \phi \qquad (27.1)$$

$$(1-\lambda)b_2x_2+b_3x_3+\lambda b_4x_4 \geqslant \phi \qquad (27.2)$$

301. Since $x_0=x_2=$ . . . . and $x_1=x_3=$ . . . a safe solution is given by

$$x_0 = \mathrm{Max}\left[\frac{\phi-b_ix_i}{(1-\lambda)b_{2j}+\lambda b_{2j+2}}\right] \qquad (28)$$

302. It is not difficult to show that:

the outer span always gives the maximum value for x_0,

beams with fixed ends yield larger x_0 than those with free ends,

the maximum value of x_0 with respect to the loading scheme corresponds to a central concentrated load when the ends are fixed, and to a pair of concentrated loads at $l/4$ when the ends are freely supported.

303. The following geometrical interpretation can be given to this solution. Relations (27) represent lines in the plane (x_0, x_i) as shown in Fig. 34. The solution is a point within the "safe domain" bounded by the lines corresponding to the imposed conditions (26.3), (26.4) and (28). The most economic design is to be found on the line (27.1). If a convenient lower bound of the span yield safety parameter is assessed (e.g. $x_1=x_3=$ . . . 0·9), the solution is the intersection point of the lines (27.1) and $x_1=0{\cdot}9$.

304. The solution in the first procedure is that indicated in Fig. 34 with a dotted line.

305. Detailed calculations performed for continuous beams with 2, 3, 4, and 5 equal spans, with free and fixed-ended supports and for 15 cases of loading have enabled the upper bounds of x_0 to be obtained. Therefrom the following simple rule for the limit design of continuous beams with equal spans was stated.[67]

Plastic moments can be derived on the basis of the corresponding maximum elastic moments (relation (21)), reducing their values for the factored live load as follows:

(1) at span critical sections by 10% in all cases;
(2) at support critical sections of beams with 2, 3, and 4 or more spans
by 20, 25, and 30% respectively for freely supported beams;
by 15, 20. and 25% respectively, for fixed-ended beams.

306. As an illustration of this rule and for comparison with other similar recommendations a five span continuous beam uniformly loaded is designed according to (I) the Soviet recommendations[68] (giving the same results for the inner support and span sections as the rule in the Report). (II) the elastic moments for the factored load; (III) the first OLD procedure; (IV) the second OLD procedure.

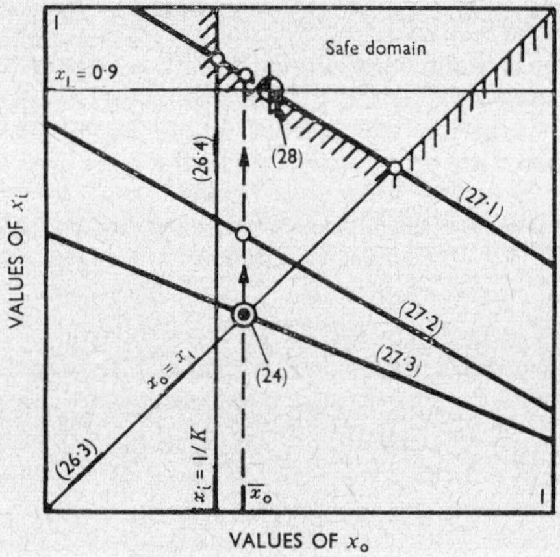

FIG. 34

307. Results are summarized in the diagram in Fig. 35 giving the x_i values for various sections to be used with the relation (21).

Rotation compatibility

308. The limit design of a structure is possible so long as the effective hinge discontinuities do not exceed the available rotation capacities of plastic hinges at any loading level. This general requirement of compatibility can be expressed as

$$\theta_i \leqslant \theta_{pi} \quad . \quad . \quad . \quad . \quad . \quad . \quad . \quad . \quad . \quad . \quad (29)$$

309. If condition (29) is fulfilled up to the factored load then this is the actual ultimate load of the given structure. If at a certain intermediate load $P < P_u < KP$ there is at least one hinge at which $\theta_i > \theta_{pi}$, then $G + P_u$ is the actual ultimate load. In some cases if special measures are taken to strengthen the plastic hinge i so that $\theta_i < \theta_{pi}$ it will be possible to reach the load KP.

310. A method to check the plastic rotations as the plastic hinges form throughout the structure has been presented by Prof. Hangan,[69] for structures with few redundancies. For general frames simplified solutions are still to be studied.

311. For continuous beams, however, the rotation compatibility of the hinges can easily be checked starting from the simplifying assumptions below:

(1) only the support critical sections are to be verified because the span sections of the latter have larger permissible rotations, and their plastic moments, as

given by proposed rule, do not differ too much from the corresponding elastic values;

(2) permissible rotations are given by the empirical formula in the Report;

(3) when the rotation of a support plastic hinge is considered the adjacent supports are assumed as ideal hinges;

(4) in the calculation of effective rotations, *EI* values of members are supposed to have a constant value, i.e. that for the point L_1 of the considered plastic hinge.

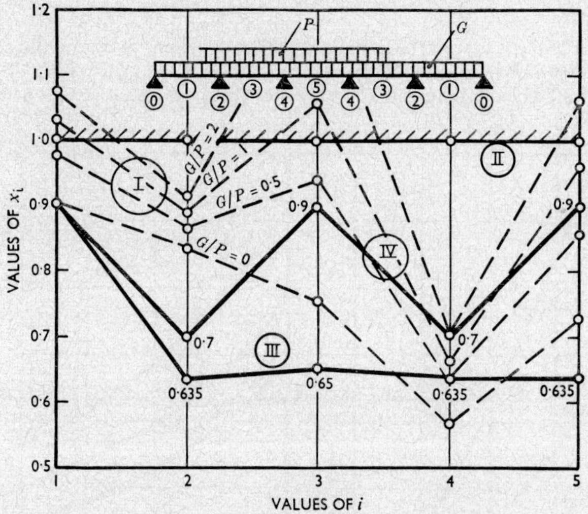

FIG. 35

312. Assumptions (3) and (4) are obviously conservative, leading to θ_i values in excess of the actual values.

313. Based on assumption (3) it is not difficult to show that θ_i may be expressed as:

$$\theta_i = \frac{\bar{X}l}{3EI}(r_i - 1) \quad . \quad . \quad . \quad . \quad . \quad . \quad . \quad . \quad (30.1)$$

or, since $\bar{X}/EI = e_{ce}/n_1 d$

$$\theta_i = \frac{e_{ce}l}{3n_1 d}(r_i - 1) \quad . \quad . \quad . \quad . \quad . \quad . \quad . \quad . \quad (30.2)$$

314. The permissible rotation being $\theta_p = 1_p(e_{cu} - e_{ce})/n_u d$ the expression of the compatibility condition (29) becomes after transformations:

$$r_i \leqslant 1 + 3\frac{l_p}{l} \cdot \frac{e_{cu} - e_{ce}}{e_{ce}} \cdot \frac{n_1}{n_u} \quad . \quad . \quad . \quad . \quad . \quad . \quad . \quad . \quad (31)$$

315. Therefore the compatibility is verified if for the corresponding critical section the redistribution factor does not exceed a definite upper bound, the value of which is given by relation (31).

316. By definition

$$r = \frac{G/KP + 1}{G/KP + x} \quad . \quad . \quad . \quad . \quad . \quad . \quad . \quad . \quad (32)$$

so that from relations (31) and (32) the compatibility appears to be dependent on the parameter G/KP, the yield safety parameter x, the relative length of the plastic zone l_p/l, the concrete strains and the depths of the compressed zones at L_1 and L_2.

317. It will be seen that the whole problem reduces to an upper limitation of the steel percentages of critical sections. To be specific consider a rectangular beam reinforced in tension only, with $c_u = 3000$ lb/sq. in. and $t_y = 40{\cdot}000$ lb/sq. in. The values of n_1 and n_u can be derived from $p = w\dfrac{E_c}{E_s}\cdot\dfrac{n_1{}^2}{1-n_1}$ and $n_u = pt_y/0{\cdot}85c_u$ respectively. Taking $w = 2/3$, $E_c = 750c_u = 2{\cdot}25 . 10^6$ lb/sq. in. and $E_s = 30 \times 10^6$ lb/sq. in., $e_{ce} = 0{\cdot}002$ and $e_{cu} = 0{\cdot}0035$ the r values from (31) can be plotted against p and l_p/l, as shown in Fig. 36. Relation (32) is represented on the same diagram giving also r in terms of x and G/KP parameters.

318. From Fig. 36 it is seen that the limit design solutions checking the compatibility requirement range inside the rectangle bounded by $x = 0{\cdot}7$ and $r = 1{\cdot}43$, and that compatibility implies only the determination of the maximum permissible p for the given problem.

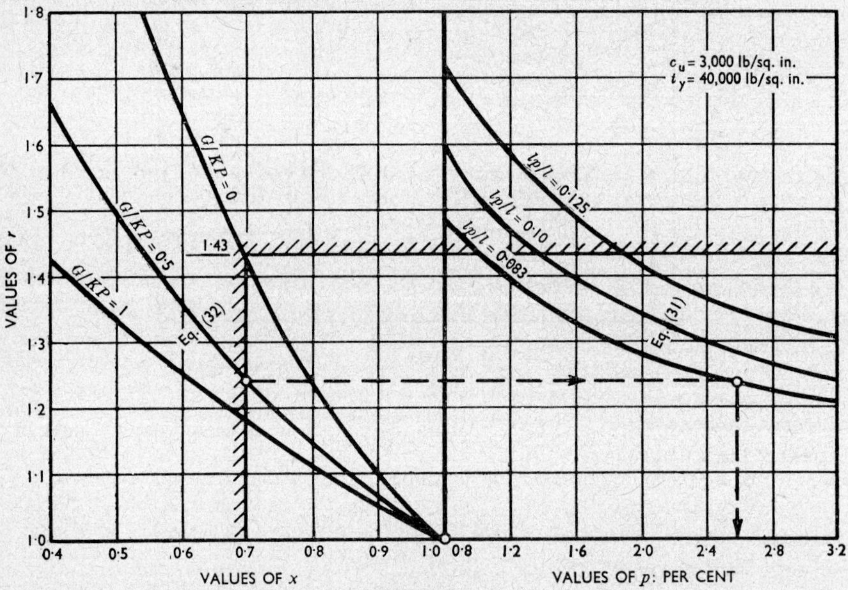

FIG. 36

319. As an example consider the limit design of a four-span continuous beam for $G/P = 1$, $K = 2$, $l/d = 12$. Taking $l_p \simeq d$ as a safe value with $x = 0{\cdot}7$, according to the rule given, $l_p/l = 12$ and $G/KP = 0{\cdot}5$, from the diagram in Fig. 36 it is concluded (see dotted line) $p_{max} = 2{\cdot}6$. Design plastic moments from relation (21) are correct if resulting steel percentages are less than this value.

320. The above results demonstrate that even with the highly conservative assumptions adopted limit design solutions are compatible up to $p = 2{\cdot}5$ or more, for current steel and concrete grades and for usual dead/live load ratios. This fact is also indicated by recent test results[70,71,72], proving full moment redistribution to occur for still larger values of p under more severe redistribution conditions.

ADDITIONAL NOTATION USED BY MR COHN

a_i, b_i = constants defining the maximum elastic moments at section i, due to the dead and live loads respectively.

b_I = $b_i x_i$

i = index referring to plastic hinge ($i=1, 2 \ldots\ldots s$)
k = index referring to a possible mechanism ($k=1, 2 \ldots\ldots m$)
m = number of elementary mechanisms
n = number of redundancies
r = redistribution factor = ratio of total loads corresponding to the failure of the structure and to the first plastic hinge respectively.
s = number of critical sections.
x_i = K_{1i}/K = yield safety parameter of plastic hinge i.
x_0 = common value of x_i for all supports of a continuous beam.
$\bar{x}_k$ = balanced value of x_i in each limit equilibrium equation.
$\bar{x}_0$ = min $\bar{x}_k$.
G = dead load.
K = ultimate load factor = ratio of ultimate to working live load.
K_{1i} = yield load factor = ratio to the working live load of the live load corresponding to the formation of a first plastic hinge at i.
P = live load.
$\bar{M}_i$ = plastic moment of critical section i.
$\bar{M}_{ok}$ = $\phi_k(G+KP)l$ = factored free bending moment for span k.
λ_{ik} = dimensionless parameter defining the location of the span plastic hinge i of the mechanism k.
w = 0·5 . . . 1 = shape factor of the compression stresses of concrete

Milík Tichý (Scientific worker, Institute of Theoretical and Applied Mechanics, Czechoslovak Academy of Sciences, Prague) wrote that sometimes appreciation of an existing structure was needed and the ultimate live load was sought. Because the structure usually carried some known dead load, only a small value of the load factor for the dead load could be considered, e.g. 1·1 instead of 1·5 as in BSCP 115:1959. The dead load mostly acted in the elastic phase of the structure, however, it consumes a certain part $\varDelta X_i$ of the ultimate moments $\bar{X}_i$ in the critical sections. For this reason the expression (4) should be transformed into a more general equation:

$$\int\frac{M_iM_0}{EI}\,ds+\sum(\bar{X}_k-\varDelta X_k)\int\frac{M_iM_k}{EI}\,ds=-\theta_i \quad . \quad . \quad . \quad . \quad . \quad (4a)$$

322. The values of the integrals did not have to be changed, because they described the elastic characteristics of the structure. Similarly, no change was needed with the values θ_i, as far as the dead load did not act in the plastic phase. However, if the dead load reached the plastic phase and a plastic rotation $\varDelta\phi_i$ due to this load occurred, it was necessary to increase the rotation ϕ_i by this value. Then, the condition

$$\theta_i+\varDelta\theta_i<\theta_{pi} \quad . \quad . \quad . \quad . \quad . \quad . \quad . \quad . \quad . \quad . \quad (33)$$

should be checked.

323. The method proposed in the Report was obviously easy for the design of many statically indeterminate structures. However, for some simple statical systems, e.g. for continuous beams, the "transfer" method allowed by BSCP 114:1957 would be more convenient from the practical point of view. The same method was now recommended in the proposed Czechoslovak Code of Practice for reinforced concrete structures (ČSN 731201) with the difference only, that the elastic support moments could be corrected by ±20% of their value (with appropriate changes in the span moments). In order to check this proposal the Institute of Theoretical and Applied Mechanics of the Czechoslovak Academy of Sciences had tested some continuous beams. The experimental results together with theoretical considerations had specified the influence of the relation between working moments (elastic) and the ultimate moments in the critical sections upon the degree of the moment redistribution. This influence had been suggested previously by Macchi[60a] for prestressed concrete beams.

324. Let us denote in a given span of a continuous beam the critical sections by

such a way that in the section 1 the smaller of the two critical section moments is acting (M_1, elastic, in absolute value). In the section 2 the absolute value of the elastic moment is M_2. The two moments acted in opposite sense (Fig. 37). The Institute had found that the condition

$$\frac{M_{ult1}}{M_1} \cdot \frac{M_2}{M_{ult2}} < 2 \quad . \quad . \quad . \quad . \quad . \quad . \quad . \quad . \quad . \qquad (34)$$

should be fulfilled, if the moment redistribution at the ultimate load should be good. The lower limit of the ratio need not to be sought, because its value follows from a logical design. For practical reasons we usually aim to equalize (as far as possible) the ultimate moments in the critical sections, so that the lower practical limit cannot be smaller than 1 (when $M_{ult1}:M_1 = M_{ult2}:M_2$).

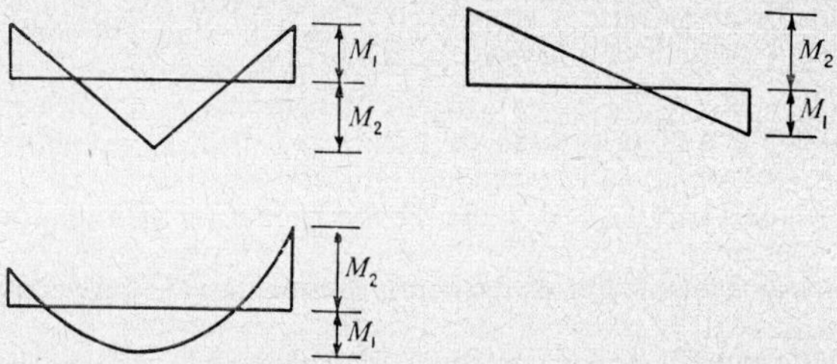

FIG. 37.—ELASTIC MOMENTS, M_1, M_2, IN A SPAN FOR DIFFERENT BENDING MOMENT DIAGRAMS

325. Consider now an inner span of the continuous beam with a symmetrical bending moments diagram (Fig. 38). Let us suppose that

$$M_1 = M_S < M_M = M_2$$

Then, according to the condition (34), it must be

$$\frac{M_{ultS}}{M_{ultM}} < 2\frac{M_S}{M_M} \quad . \quad . \quad . \quad . \quad . \quad . \quad . \quad . \quad . \qquad (34a)$$

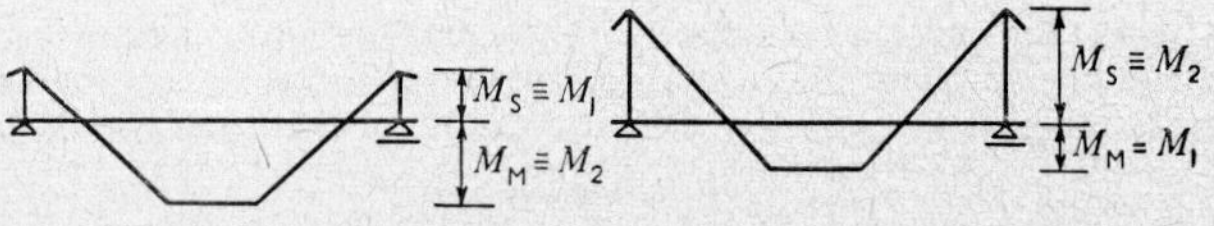

FIG. 38.—ELASTIC MOMENTS, M_S, M_M, IN AN INTERNAL SPAN OF A CONTINUOUS BEAM

326. The value, by which we are *increasing* the support moment, will be ϕM_S, and the new design moments are as follows:

$$M_S^* = M_S + \phi M_S,$$
$$M_M^* = M_M - \phi M_S.$$

327. Obviously, the structure will be designed in such a way that it will be

$$\frac{M_S^*}{M_M^*} = \frac{M_{ultS}}{M_{ultM}}.$$

After putting into the equation (34a), we obtain

$$\frac{M_S^*}{M_M^*} < 2\frac{M_S}{M_M}$$

and further

$$\frac{M_S+\phi M_S}{M_M-\phi M_S}<2\frac{M_S}{M_M}.$$

328. Finally, after some operations, it must be (for $M_S/M_M<1$)

$$\phi<\frac{1}{1+2(M_S/M_M)}.$$

329. Analogously, if it is $M_1=M_M<M_S=M_2$ (i.e. $M_S/M_M>1$), we must get from (34) the condition

$$\phi<\frac{1}{2+M_S/M_M}$$

330. The maximum values of ϕ are traced in the Fig. 39 as a function of the ratio M_S/M_M. We must consider only, that for $M_S/M_M<1$ the ϕ is additive, whereas for M_S/M_M it is subtractive.

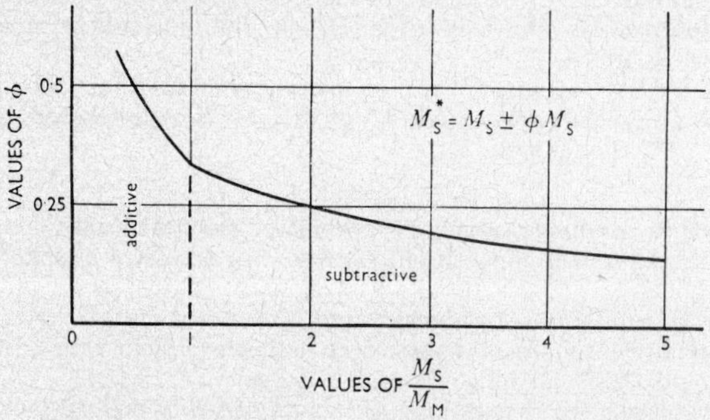

FIG. 39.—VARIATION OF ϕ WITH THE M_S/M_M RATIO

331. From Fig. 39 we can find that the 15% allowance of the BSCP 114 is safe up to the ratio $M_S/M_M=4{\cdot}1$, and the Czechoslovak Standard (with $\phi=\pm 20\%$) up to $M_S/M_M=3$. In both prescriptions cases with $M_S/M_M<1$ are underestimated, being there possible to permit a correction greater than 0·33. Regarding to the fact that in practice a ratio $M_S/M_M>2{\cdot}5$ does not occur, a correction by 0, 25 should be allowed. This fact has been proved in a different way by Mattock[73].

331. Mr Tichý ended his contribution by affirming that the Report rendered good instruction on the design of statically indeterminate structures. Some points could be omitted, since they had more general significance, e.g. the determination of the load factor. On the other hand the Report should be completed by some information about the design of structures with varying moments of inertia, especially concerning the choice of positions of hinges in such structures.

332. Mr Tichý thought it not without interest that the latest edition of the Soviet instruction for the design of reinforced and prestressed concrete statically indeterminate structures[74] recommended the use of equation (5) for precast prestressed concrete structures. For monolithic structures the moment redistribution method (limit design) under certain conditional assumptions is used.

Mr W. J. R. Smyth (Senior Engineer, Ove Arup & Partners) wrote that most of the Report was devoted to the description of two methods of analysis, one for frames and one for slabs; and it was a pity that there was not more discussion of basic principles or of the general theorems which had been put forward elsewhere. The idea of upper and lower bounds was not mentioned (except for the use of a limiting stress/strain diagram), in spite of both these methods giving, generally, upper bounds for the yield moments which might therefore be unsafe solutions unless checked by other means (by drawing the bending moment diagram for frames, and for slabs by finding a lower bound to the moment or by empirical knowledge of the worst patterns of yield lines). These points had been discussed by J. F. Baker and others[51] in connexion with steel frames and by Wood[76,77] for slabs.

334. The two main objections to elastic analysis were firstly the lack of correspondence between the model analysed and the actual structure, and secondly the amount of calculation involved, which often led to alternative loading conditions not being properly considered. Both of these objections applied almost equally to the calculation of the rotations of yield hinges as described in the Report.

335. The examples given for calculation of hinge rotations did not show what they purported to show. In example 1, using the figures given, the allowable support moments would lie between the full value of the simply supported moment, and one sixth of it. Near either of these extremes one would expect to have bad cracking under working loads—in one case in the centre of the span, and in the other at the supports. In this case an approximate knowledge of the elastic solution would be more useful than the calculation of rotations.

336. In example 3 the moment at the outer corner had no doubt been ill-chosen on purpose to show an excessive rotation at hinge lb. Mr Smyth believed, however, that the equation in § 151 should read

$$-2\theta_{1b} = \bar{X}_{1b}\delta_{1b \cdot 1b} + \ldots$$

since the rotation in two hinges was being considered simultaneously.

337. This meant that the hinge rotations were not excessive in spite of the badly chosen corner moment.

338. The hinge rotations in examples 1 and 3 were not critical, and in both these cases values of redundant moments chosen by a designer familiar with elastic behaviour of structures would satisfy the hinge-rotation criterion.

339. Mr Smyth believed that as long as factors of safety or load factors remained at their present level, it should not be necessary in any normal structure to calculate hinge rotations, but that under working load the structure should not depart too far from an elastic structure—the Danish recommendations in § 124 seemed to attempt to ensure this.

340. The method given in § 66 for deciding the load factor seemed to be based on a series of arbitrary numbers, and two different persons using it would produce two different factors. If load factors could not be assessed on an entirely statistical basis the recommendations of the European Committee on Concrete seemed to be less arbitrary and simpler to use; also they made allowance for varying standards of calculation, and this was quite important in practice.

Mr R. J. Dailey (Partner, A. B. Dailey, Son & Clarke, Consulting Engineers, London) hoped that the labour of studying the Committee's present and future proposals would be repaid by a simplification of present office design procedure in the near future. The examples quoted of Danish and Russian practice were very attractive.

342. For any but the simplest structure the procedure appeared to have big pitfalls for the unwary. He hoped that the proposals would not be introduced into Codes of Practice in an obligatory form at first, but in the same permissive way in which the yield-line theory for slabs was introduced into C.P. 114:1957. This would enable designers to get the feel of the method.

343. In § 68*c* there was a reference to the effect of settlement of supports, and he felt that this was a very important matter. Safe limiting values of the likely amount of relative settlement should be chosen. Then the hinge rotation it produced could be deducted from the total permissible rotation, leaving the amount of rotation available

under the loading cases. Terzaghi and Peck[78] suggested ¾ in. as the limiting value of relative settlement between adjoining columns. With column centres of 15 ft this gave a hinge rotation of 0·0042 radian. If alternate columns settled by this amount the total rotation in the hinge adjoining intermediate columns would be 0·0084 radian. In Example 1, § 135, the total permissible rotation was only 0·01. Thus the rotation due to settlement could be perilously close to the total permissible rotation. Relative settlements normally occurred slowly, and if θ_p increased with age the position might not be so serious. However there might be a critical age at which the load factor was at a minimum, as illustrated in Fig. 40. Could the committee give any information on the rate of increase of θ_p with age?

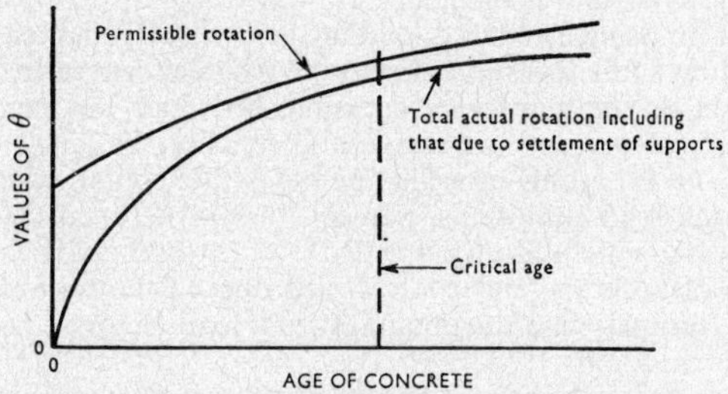

FIG. 40.—POSSIBLE RELATIONSHIP BETWEEN PERMISSIBLE AND ACTUAL ROTATION OF HINGES AT INCREASING AGES, INCLUDING THE EFFECT OF RELATIVE SETTLEMENT

344. A further point concerned fatigue effects, referred to in § 26. Tests by the Building Research Station[79] showed that loads of only 0·7 times the ultimate static load would just not cause failure of reinforced concrete beams if repeated one million times. Practical structures were unlikely to have to endure this number of repetitions of load, but the effect appeared sufficiently important to warrant inclusion in the load factor calculation of § 66. Was it intended to cover this in parameter (*f*)?

345. The durability of the concrete at exposed hinges was a point of concern. Presumably the crack widths would be greater at hinges than elsewhere. Little research had yet been published on the effect on durability of the crack widths, and it was probably desirable to avoid the occurrence of hinges in exposed positions. This would add to the complexity of the design procedure.

346. Mr Dailey noted from Eq. (8), § 50, that the value of n_u was important in the determination of θ_p, but felt that insufficient guidance was given as to its determination. Table 3 laid down that it was to be found from the relationship $C=T$, but did not stipulate a safe limiting value for α. If α was taken as 0·67, from Table 1, then with $c_u=3000$ lb/sq. in., $t_u=80{,}000$ lb/sq. in. at L_1 and with 1·2% of steel, $n_u=0{\cdot}56$. However if one worked from the linear strain relationship (§ 53a), when $e_{cu}=0{\cdot}0035$ and $e_{su}=$ say 0·015 then $n_u=0{\cdot}19$. Differences of this order would clearly have a big influence on the value of θ_p, and Mr Dailey would be glad of clarification on the matter.

Professor M. R. Horne (Professor of Civil Engineering, the University of Manchester) wrote that in the Report, he was recorded as being in disagreement with the clauses on load factor that allowed for a reduction down to a value of 1·25 in certain cases (§ 65). This important section of the Report as written had, in his opinion, no sound basis whatsoever, and he opposed the suggestions made in §§ 64 and 66 to 70.

348. The proposals were based on one of the alternative schemes presented in the "Report on Safety" of the Institution of Structural Engineers.[80] In that report, various means were explored of arriving at reasonable load or safety factors in the design of structures, the proposal adopted by the present committee being closely

similar to those contained in Appendix II of the reference. While the Report on Structural Safety had deservedly been welcomed, Professor Horne had always understood that the numerical factors suggested were intended as merely illustrative of the methods proposed. The Report on Safety did in fact state that more extensive study would be required to establish any of the suggested schemes.

349. The most questionable numerical factor would seem to be the parameter for "Load conditions, including consideration given to combined loads" (§ 66, parameter (c)). The variation of certainty over load intensity was allowed for by an absolute variation of 0·125 in the load factor, corresponding to a 10% variation at the minimum load factor of 1·25 and to a 6·25% variation at the maximum load factor of 2·00. In view of the widely varying conditions under which loads had to be specified, this degree of adjustment appeared quite inadequate. It was true that, in § 68, one of the conditions to be satisfied in using a reduced load factor was that "the loads should be maximum values based on statistical surveys or some reliable form of load control." Were this strictly adhered to, the number of occasions on which the formula could be used would, on the basis of present knowledge and practice, be few. Unfortunately, the force of this proviso was entirely cancelled by the phrase contained in § 66(*c*) on the use of the parameter for load conditions, namely "assess between 0 (accurate and controlled) and 4 (less accurate and uncontrolled)." With no guidance as to what might constitute degrees of accuracy and control, this opened the way to an indiscriminate application of the formula, with the possibility of a load factor as low as 1·375 even for ill-defined loading conditions.

350. Apart from the fact that the parameters suggested had been derived by pure guesswork, the most serious feature of the arguments used in the Report was that they suggested that a load factor of 2·00 would to be entirely adequate under any circumstances. Taking the most unfavourable conditions mentioned in § 66, a load factor of 2·00 was deemed satisfactory for the design of a vital member of a structure of which the results of failure would be very serious with no warning, the workmanship being occasionally bad and no maintenance likely. Associated with all this would be an

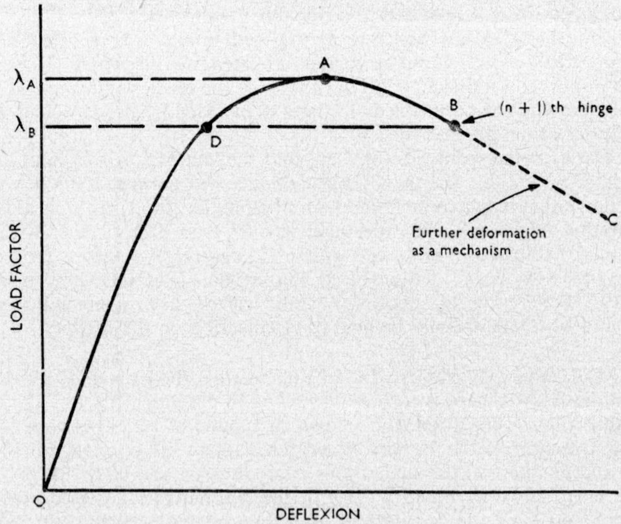

FIG. 41. LOAD/DEFLEXION CURVE FOR ELASTIC–PLASTIC STRUCTURE SUBJECT TO INSTABILITY

assessment of loads to be regarded as appreciably less than ideal, with no form at all of load control. It was hard to believe that any engineer concerned with his integrity and reputation would risk a load factor of 2·00 under these conditions, and yet this was specifically recommended in the Report.

351. In making these criticisms, Professor Horne did not suggest that the load factor should not, under favourable conditions, be less than 2·00. He did however maintain that the proposals set forth represented an inadequate and possibly dangerous procedure.

352. The mention of frame instability in § 83 was inadequate, and gave the wrong impression. The important point was that the design method stopped one hinge short of a complete collapse mechanism. Hence the deflexions throughout the structure under the specified loads were known, and their effect on carrying capacity was allowed for as described in §§ 74 to 81. This did not necessarily mean that the structure would be stable under the state of "n" plastic hinges that had been assumed in the design. The condition calculated was immediately prior to the formation of the $(n+1)$th hinge which transformed the structure into a mechanism, corresponding to point B on the load/deflexion curve, Fig. 41. In a multi-storey structure, or perhaps in a lightly loaded and therefore slender, single-storey frame, the point B might be well advanced along the drooping portion AC of the curve, denoting a condition of instability. The failure load factor λ_A could not however be lower than the design load factor λ_B, and the real condition of the structure when loaded to intensity λ_B would be represented by point D. Although the corresponding state of the structure was not calculated, it was known that it must exist, and that D was on the stable, rising portion OA of the load-deflexion curve. This was the justification for saying that frame instability was of no significance in the proposed design method—the effect of frame instability on carrying capacity was in fact allowed for. A similar procedure could be adopted for steel frames, and it should not be imagined that concrete frames were by nature immune from the instability problems that had been discussed elsewhere in relation to steel frames.

Dr E. W. Bennett (Senior Lecturer in Civil Engineering, University of Leeds) wrote that in recent studies at the University of Leeds the available results of tests on continuous prestressed concrete beams had been compared with several theories. The "simplified limit method" of the Report, in its original form proposed by Professor Baker[81], was found disappointing in that it almost always grossly underestimated the experimental strength of the beams. This appeared to result partly from an insufficient allowance for the rotation angles, which had been rectified by the introduction of the empirical formula in the Report. The method might still, however, be criticized for introducing only sufficient hinges to render the structure statically determinate, and then assuming that the deformation of the members between these hinges could be represented by a constant flexural rigidity (EI). The "high local increases of curvature" referred to in § 28 were thus ignored *between* the hinges. The curvature here, however, could be so great as to amount to an additional hinge (as in the plastic theory of steel structures), especially under a concentrated load, even though the ultimate moment might not be attained.

354. The inadequacy of the concept of constant flexural rigidity had been experimentally demonstrated not only by Dr Bennett's own tests but also by some tests on simply supported prestressed beams by Macchi[82] in which the rotation measured over a length of 10 in. under a concentrated load at the centre of a span of 9 ft 3 in. was compared with the total rotation over a length of 40 in. A calculation based on a constant flexural rigidity indicated that only 29% of the total rotation over the 40-in. length should occur in the middle 10 in.; experimentally the proportion was about 50% at 70% of the ultimate load and 65% at failure.

355. It would, therefore, seem that a more accurate physical representation of the deformed structure might be obtained by introducing sufficient hinges to convert the structure to a mechanism, as in the plastic analysis of steel structures. The rotation

permitted at each hinge must, however, be restricted because of the limited ductility of reinforced or prestressed concrete members, and the simplest assumption was that there was a unique relationship between the moment of resistance and angle of rotation, both expressed as fractions of their ultimate values, for all hinges. On the basis of the available experimental evidence the relationship $\frac{M}{M_u}=0{\cdot}4+0{\cdot}6\frac{\theta}{\theta_u}$ had been tentatively suggested for prestressed concrete.[83]

356. The use of this method of analysis was extremely simple since once the hinges had been located the value of $\frac{\theta}{\theta_u}$ was readily obtained from the geometry of the deformed structure. For example in the continuous beam *ABA′* of two equal spans with central concentrated loads shown in Fig. 42a, $\theta_P=\theta_B$, so that if the ultimate moment M_{Bu} and rotation θ_{Bu} occurred at *B*, the maximum moment at *P* (or *P′*) was given by $\frac{M_P}{M_{Pu}}=0{\cdot}4+0{\cdot}6\times1{\cdot}0=1{\cdot}0$. The hypothesis therefore indicated a complete redistribution of moment with simultaneous failure at *B* and at *P* (or *P′*). This was in fact the conclusion

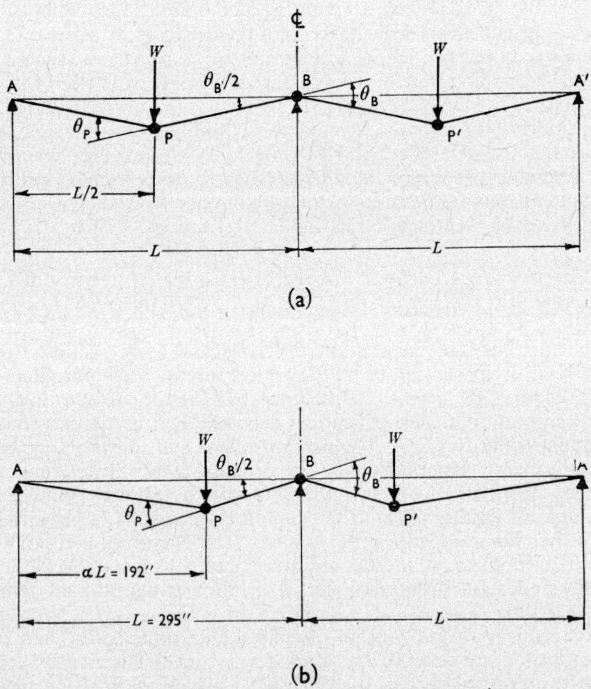

FIG. 42.—IDEALIZED ULTIMATE DEFORMATION OF PRESTRESSED CONTINUOUS BEAMS

of Morice and Lewis[84] from a series of 28 tests of continuous prestressed beams loaded in this manner, introducing a wide range of prestress and reinforcement conditions at the critical sections.

357. More generally, when the loads are applied at a distance αL from the outer supports, as in Fig. 42b, $\theta_P=\frac{\theta_B}{2\alpha}$. The numbers in Fig. 42b referred to a test reported by

Lin[85] and here $\theta_P = \frac{\theta_B}{2 \times 0{\cdot}65} = 0{\cdot}77\theta_B$. If $\theta_B = \theta_{Bu}$, $\theta_P = 0{\cdot}77\theta_{Pu}$ and $\frac{M_P}{M_{Pu}} = 0{\cdot}4 + 0{\cdot}6 \times 0{\cdot}77 = 0{\cdot}862$. Since the calculated values of M_{Bu} and M_{Pu}, according to BSCP 115, were respectively 1692 and 1330 lb-in. × 1000, the calculated ultimate load on the two spans was therefore 67,000 lb., compared with the experimental value of 78,400. The discrepancy was almost entirely due to the calculated ultimate moments, and if the experimental values of the ultimate moments were used, there was agreement to within 2%. The method proposed in the Report gave about the same value for a calculated value of M_{Bu} and about 72,800 lb. using the experimental value.

358. The above method had so far only been applied to the available test data for continuous prestressed beams but these results were extremely promising and it was suggested that a similar approach, with suitable modifications, might be used for other types of indeterminate structure. It had the great advantage of simplicity and did not demand the introduction of a large number of doubtful parameters, many of which were notoriously variable and some of which probably had little influence on the final result.

Dr R. H. Wood (Senior Principal Scientific Officer, D.S.I.R. Building Research Station) wrote that considerable research had been carried out in recent years on multi-storey frameworks in the elasto-plastic range and on the collapse of slabs, and these two aspects of design formed a large proportion of the committee's recommendations. In proposing a design method which provided a complete bending moment diagram for the whole frame, the committee clearly intended to follow a "safe" —or "lower-bound"[49] approach—which was praiseworthy insofar as many attempts at ultimate load design had to be satisfied with an "unsafe" or "upper-bound" approach by way of complete mechanisms. Whilst an elastic analysis was still necessary on account of limiting hinge rotations, this provided valuable additional information as to deflexions, and was carried out on a structure made statically determinate by a choice of impending mechanism of collapse requiring just one more plastic hinge per storey. Whereas this had been formulated in terms of the Müller–Breslau (or "influence coefficients" or "flexibility") approach[87], most designers preferred the more direct "stiffness" approach which, via the slope-deflexion equations or the Hardy-Cross technique, was also a more powerful method where problems of instability were involved[88]. Dr Wood considered that the committee's recommendations would gain in power if they were also formulated so as to allow moment-distribution methods.

360. Dr Wood had three main criticisms of the Report, which dealt with the following aspects of the subject:

(*a*) the attempt to apply, to *extensive* mechanisms of collapse, *idealized* plastic theory in materials which might depart considerably from that concept;

(*b*) frame instability in the plastic range, i.e. sudden collapse of the whole frame without the formation of a complete mechanism;

(*c*) the misleading "nodal-force" theory (or "equilibrium methods") when applied to plastic design of slabs[77].

361. Upper- and lower-bound theorems had grown up on the assumption of plastic moment/curvature relationships with ideally constant yield moments. Since the history of loading was not traced in such an analysis, the only guarantee that the structure would ever get into the state which was claimed (and not a different system of hinges) lay in this idealization. Almost no-one had studied what happened to the moment redistribution process when the moment/curvature diagram departed from this ideal (though a study of such generalized plasticity applied to frameworks was in progress at the Building Research Station). Thus the committee's lower limit L_1 should be treated just as seriously as L_2. When the actual moment/curvature diagram was not flat-topped, then the hinge points had ill-defined moments; if there were a large number

of (now so-called) "hinges" in the mechanism, then, when the hinge which had undergone the most strain deformed beyond the point of maximum moment, all further frame analysis became chaotic[90]. The characteristics shown in Figs 3 and 9, for cold-worked and prestressed steel, were by no means sufficiently ideal to be applied to extensive multiple-hinge mechanisms in multi-storey buildings.

362. The most important yardstick, with which to measure the likelihood of instability, was the Euler load, $P_E = \pi^2 EI/L^2$, when taken in conjunction with beam/column stiffness ratios and sidesway. Before going further it was important to see what ratios of the *factored* direct load to the Euler load, viz. P/P_E, were allowed in columns designed to BSCP 114. Thus with a 1:2:4 concrete, and 760 lb./sq. in. working stress, a modular ratio of 15, a load factor of 2·0, a square column of side d and length L would give the values shown in Table 14 for permissible values of P/P_E according to Table 18 of the Code in terms of the fully effective concrete section:

TABLE 14.—RATIOS OF FACTORED DIRECT LOAD TO EULER LOAD

L/d	P/P_E	P/P_E if effective length is allowed to be ·75L
15	·20	0·36
21	·32	0·57
27	·40	0·71
33	·40	0·71
45	·37	0·66

These figures were reduced by some 8–20% on account of the reinforcement, they were also effectively decreased when bending stresses had to be allowed for, and increased for higher grade concrete and for prestressed concrete. A typical example, given by by Prof. A. L. L. Baker[91], worked out at $P/P_E = 0·29$ to 0·38 depending on the value chosen for the modular ratio. A figure of $P/P_E = 0·1$ to 0·3 was probably representative of common practice.

363. Now when side-sway was prevented, by bracing etc., the elastic critical load was in the range $P_{crit} = 2·0$ to $3·5P_E$, which was what was meant when we spoke of an effective length l equal to $0·7L$ or less. The most detrimental effect that plasticity in the beams can then have[92] was to reduce the elastic critical load to the Euler load, so there is no danger whatever of the ordinary Euler-type buckling of individual columns (a limited form of frame instability).

364. It was therefore very curious that the committee should deal at some length with this unlikely kind of instability but considered that it could "avoid the (sway) type of (overall frame) instability observed by Wood in steel structures". For, considering first a pinned-base single-bay elastic portal frame with an infinitely stiff beam, the critical load was clearly $P_{crit} = 0·25P_E$ (i.e. $l = 2·0L$ effectively). But if one column had hinges both top and bottom (following the proposals of Fig. 13), and where plastic hinges were in fact treated as *real* hinges with equal and opposite externally applied moments, then the critical load would be expected to be about halved, i.e. $0·125P_E$ approximately (actually $0·138P_E$). This was the *top-limit* for possible deteriorated critical loads (per column) when the committee's method was used. The general case was listed in Table 15, expressed as average values of P/P_E in any storey.

365. These values were alarmingly small even when compared with practical values. They implied that a frame would collapse, suddenly and without warning, when fewer hinges than postulated had formed. But that was not all. The actual deflexion was likely to be that calculated (ignoring instability) multiplied by $\left(\frac{1}{1-P/P_{crit}}\right)$. Thus if we could not allow more than 25% further increase in hinge rotation due to this cause, then

$\frac{1}{1-P/P_{crit}}=1{\cdot}25$, or P/P_{crit} must not exceed 0·2. Consequently all the limiting values of P_{crit} in Table 15 must further be divided by 5·0 if the calculated limiting hinge rotations were to be believed. Thus the actual answer for the analytical process undertaken could be grossly in error unless frame instability was included.

366. This made it clear that the recommended design calculations were invalid and misleading unless the overturning moment ($P\times$ deflexion) was included at every joint.

TABLE 15.—AVERAGE VALUES OF P_{crit}/P_E PER STOREY WHICH WOULD PROMOTE FRAME INSTABILITY, USING THE COMMITTEE'S PROPOSED DESIGN METHOD

Number of bays	Ratio of beam to column stiffness					
	$\frac{1}{4}$	$\frac{1}{2}$	1	2	4	∞
1	·03	·050	·074	·096	·113	·138
2	·02	·033	·050	·066	·078	·095
4	·012	·020	·030	·040	·048	·058
8	·007	·011	·017	·022	·026	·033
16	·0035	·0065	·0095	·012	·015	·020

Even then convergence would be a problem and might well be impossible with so many hinges. It would appear that:

(1) sway should be avoided if at all possible;

(2) instead of a dubious, extensive and impractical lower-bound analysis*, a more practical method was to raise the load factor (about 25%) to guard against frame instability and then to use a design based on a simple complete mechanism: a simple approach such as this was justifiable in view of the help from composite action and its effect on the real distribution of loads.

It must be remembered that when limiting concrete strains were the criterion for acceptance of hinge rotations, frame instability must be taken just as seriously as with steel structures, which, on the contrary, derived some help from strain hardening.

367. It was unfortunate that (§§ 105–110) the committee should have given the nodal-force theory such a prominent position in the design of slabs. In allowing the work equations as an alternative, "... in practice ... often used", this made it look as though the shear forces (=nodal forces) in the slab *must* be taken into account. This would certainly confuse the designer especially as it was now known[89] that nodal-force theory was not always correct. Thus the most controversial aspect of slab design was the one given most prominence in the committee's proposals. Far from being compulsory, it should be allowed only as an aid to design. It was better to admit frankly that slab design at the moment was based on the upper-bound technique only (Hillerborg's method excepted). The designer should be expected to check alternative modes of collapse.

Dr R. Gartner (Consulting Civil Engineer, Cape Town) wrote that it had been known for many years that the design of reinforced concrete structures based on the elastic properties, gave results which were safe, but that the safety factor differed considerably, whilst ultimate load design gave more consistent values.

368. The time was now ripe therefore that recommendations for ultimate load

* Except by computers. The B.R.S. programme SPRORF (Stability in the Plastic Range of Rectangular Frameworks), copes with this problem on the Ferranti Mercury computer.

design of reinforced concrete structures should be made. This was a formidable task. Of the two main materials in reinforced concrete, one itself consisted of four different ingredients, namely cement, stone, sand, and water.

369. The properties of the sand and stone differed not only from town to town but might even differ from job to job. In addition the mixing proportions of these ingredients might vary, so too could the preparation. It was therefore not surprising that the stress/strain relationship for concrete varied considerably. The proposal to use an idealized stress/strain curve with two distinct points *I*1 and *L*2 was therefore to be recommended.

370. Coming to the Report itself, several points could still be investigated of which the following might be mentioned.

371. In the Report it was suggested that the elastic solution should be calculated at least approximately, but there were cases where the elastic solution only should be used. These were mainly the cases in which structures must not only stand up to the imposed loads, but should also develop no cracks at working load, such as tanks, reservoirs, structures exposed to sea water or corrosion. Also if structures were exposed to high temperatures, or variations of temperature, the ultimate load design should not be applied.

372. Another point was that in this Report the deformation due to bending alone was considered. Even if this part of the deformation was the most important, the deformation from other causes should also sometimes be investigated. For example:

(*a*) Direct compression in shallow arches, frames, or columns especially heavily loaded.
(*b*) The elongation of ties.
(*c*) Shear; apart from torsional shear, this deformation was generally very small.
(*d*) Temperature. The deflexion or rotation caused by normal temperature should always be included in the equations for the unknowns.

373. On page 407 the title for § 28 was "Rigorous analysis". In this paragraph it was pointed out that such a rigorous analysis was doubtful, therefore the title could better be changed to "Elastic analysis".

374. The next paragraph, 29, again had the title "Rigorous analysis". Here the word "Rigorous" also seemed to be too strong, and could be omitted or amended.

375. The equations for the unknowns were given in this Report, numbered (1) to (4). According to these equations, not only the elastic rotations, but also the plastic ones, were carried over from one end of the frame (Fig. 5) to the other. This carrying over of the plastic rotations seemed to be very doubtful. Tests on a small scale would be very desirable to settle this point.

376. *Example* 3. Page 434, Case 1, Vertical loads. Here six hinges were inserted whereas there were only five unknowns. As long as the loading was symmetrical, as it was in this case, this system was stable, but the least horizontal or unsymmetrical loading would make it unstable. Such a choice, of the positions of the hinges, seemed to be rather unfortunate, especially as in Case 2, where another statically determinate system, which was perfectly stable, had to be used. It would have been just as well to use the same system in both cases. Furthermore a plastic hinge in the middle of the span should be avoided if possible. The reason for this was: A plastic hinge in the middle of an overloaded structure caused excessive deflexion and sudden collapse, whereas a hinge over the support showed ample distress signs at overloading. Furthermore the equations were rather sensitive with widely differing δ_{ik}-values. In this example the corner moment had, in the determinate case, the value 10×10^6 and ended up with $1{\cdot}1 \times 10^6$. It was clear that small changes of this value gave big differences and the calculation must therefore be much more exact.

377. Dr Gartner ended his comment with the remark that the Report was a great step forward in the "Ultimate Load Design of Concrete Structures", and would be a great help, if this sort of calculation could be extended to beams with shear-failure.

Dr Geoffrey C. Brock (Head of Department of Civil Engineering, Loughborough College of Technology) wrote that engineers discussing ultimate load philosophy during recent years had often overlooked that it guided the work of several of the pioneer constructors in reinforced concrete. Through their experience of ad hoc tests to destruction they learned a good deal about the conditions which accompany failure. It was only the influence of the various national commissions, established during the period 1904–1909, which compelled this approach to be abandoned in favour of calculations based on stresses at working loads. But there was by no means unanimous agreement on the wisdom of imposing elastic theory on reinforced concrete. According to E. Freyssinet all the affirmations of the French Commission on Reinforced Concrete of 1906 were radically false, and both Rabut and Considère refused to be associated with them[93]. Commenting upon elastic theory in USA, L. J. Mensch wrote "The assumptions recommended as a basis for calculation . . . will hinder the clear understanding of the mechanics of reinforced concrete for another ten years or more. . . . Formulas based on the conditions prevailing at the time of the ultimate load are considerably simpler and more correct, because they agree with facts".[94]

379. It was possible to see now that in 1918 Mensch seriously underestimated the number of years during which unsuitable regulations could impede progress. It was important that engineers at the present time should recognize the danger that premature regulations concerning ultimate load methods could prove equally unfortunate, especially if imposed on an international basis. For this reason any report, such as the one now under discussion, which might possibly lead to new and restrictive regulations should be subject to the severest possible scrutiny.

380. When the Report is viewed in this light, two main criticisms presented themselves,

(1) The methods presented would be more properly described as those for analysis rather than design

(2) There had been no attempt to collect the experimental justification which was necessary for the acceptance of such methods.

381. The first objection would doubtless represent the reaction of a number of design engineers. They would ask whether the complicated processes suggested for checking the compatibility of rotations in an elasto-plastic frame would be of assistance in creating a structure. Many of them would probably lean towards the point of view recently expressed by Riccardo Morandi, who questioned whether calculation could be considered an absolute factor in the determination of a structure[95]. Design must always be a positive process in which the engineer takes decisions to shape a structure to his will: analysis is a passive one where at best he makes a detached scientific assessment and at worst follows through a technical ritual.

382. The rigorous analysis of plane frames, which occupied over half of the Report, could only infrequently be justified in practice. Frames usually occurred as parts of three-dimensional structures which were stiffened by monolithic slabs. If, for the purpose of analysis, the slabs were ignored or assumed to be cut on a line parallel to the frame, a considerable simplification was being introduced. It was difficult to reconcile the careful check on rotations with such simplifications. Furthermore, the calculation of these rotations was based upon the premise that plane sections remained plane after bending. While this might be true at sections of constant bending moment, such as those where the moment/rotation characteristics are determined, experiments showed that it did not hold where there were important shearing forces.

383. For the design process it seemed preferable to assume that the fully plastic collapse load of a frame could be reached and to take the necessary precautions in detailing to ensure that this would be so. Dr Brock had analysed the results from 52 two-span beams tested at the University of Illinois[96]. Thirty-seven of these beams reached collapse loads in good agreement with those predicted by simple plastic theory: the others had no web reinforcement. It was not clear how the committee reached the conclusion

that a deformation check would be necessary, but it seemed possible that they were influenced by tests on poorly detailed beams. There was a very real need for experimental evidence on the effectiveness of different reinforcing details in realizing the fully plastic potential of frames. Such tests would include shear, bond and anchorage effects, which were all controlled by detailing. But it would be wrong to expect that from such tests we would be able to produce a set of rules to replace an imaginative detailer.

384. It was suggested in the section of the Report which deals with yield-line methods that they had found complete experimental justification. But most of the tests had been on relatively simple configurations. When more complex, and more typical, structures had been tested the importance of membrane action has been revealed.[97] Designers would need to exploit the increase of strength which arose from confining the boundaries of slabs and for which further investigations were needed.

385. Finally, the Report seemed based upon the assumption that design would always be the product of calculation. Cheap and rapid model techniques were being developed for determining the collapse behaviour of reinforced structures and an indication of one way in which they might develop had been given by C. Benito[98]. They might well become a more fruitful source of information and stimulus for the designer than increasingly complex schematic computations.

Dr K. Poologasundram (Designs and Research Engineer, Ceylon Cement Corporation) commented first on the uniqueness of the final state of collapse.

387. Consider a statically indeterminate structure, subjected to increasing loads, which are proportional, in the ratio λ to the working loads. If, for a certain value of λ, the elastic limit is reached in one section, the distribution of bending moments generally begins to be modified; the internal stresses in a given section cease to vary linearly with λ.

388. At a section A, where the elastic limit is reached, the rate of increase of bending moment decreases and, consequently, causes an acceleration of the increase of the moment at other sections B, C, D etc., this may be represented graphically as shown in Fig. 43.

389. For any value of λ greater than λ_A the additional load corresponding to the increase in λ does not result in any increase in the moment at section A. During this increase in load, the structure behaves as though section A acts as a hinge. It is known as a plastic hinge, meaning that so far as the load is concerned beyond λ_A it acts as a hinge exerting a constant and known bending moment equal to M_{pA}, the resisting moment of section A.

390. As the load is further increased, section B will in its turn reach the plastic state for a certain value λ equal to λ_B (Fig. 43). Similarly section C will in due course reach the plastic stage for λ equal to λ_C and so on for other sections. The schematic shape of moment/load curves for section A, B, C etc. is equivalent to the hypothesis of successive elastic behaviours with different stiffnesses of the structure.

391. If the structure taken as a whole has n redundancies, the number of plastic hinges which can appear without the structure becoming unstable is n and the formation of the $(n+1)$th hinge results in failure. If the load factor and the final state of the structure are the only aims, it is not necessary to worry about the successive states of hinge formation. A knowledge of the location of the hinges and of the value of the plastic moments is sufficient to determine the final state.

392. The method of distributing the bending moments at the final state and establishing the uniqueness of that state has been forwarded and proved[32]. This will be apparent below.

The uniqueness theorem

393. If, for a framework n times statically indeterminate and subjected to proportional loading,

(*a*) a bending moment distribution is found, which is in euqilibrium with the given loads corresponding to a load factor λ, and is such that the bending moment at n cross-sections which do not correspond to a mechanism of failure are equal to the ultimate moment, and at every other section is less than the ultimate moment,

(*b*) the rotations at the above sections are positive.

Then the load factor λ will be less than the actual collapse load factor λ_c and plastic hinges will form at the above-mentioned n sections and in no other section.

394. Fig. 44 shows a distribution of bending moments in a three-bay, three-storey frame, satisfying the uniqueness theorem.

Rotation of plastic hinges

395. The assumption that the moment at a hinge remains constant while the moments increase in the rest of the structure implies a rotation of this hinge. In fact,

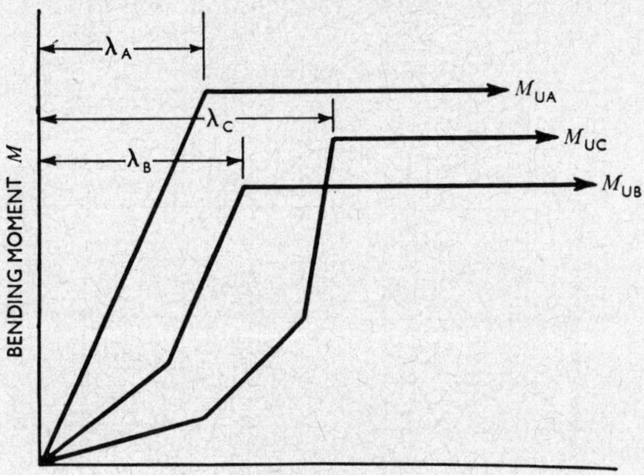

FIG. 43

however, since a hinge is not located at a section alone (i.e. on an infinitesimally small length) but is distributed over a certain length which has reached a plastic state, the rotation of the hinge is the sum of the elements of curvature $\left(\frac{1}{\mathrm{R}}\right)$; in other words if θ is the rotation which is required at a hinge to permit full redistribution of moments then $\theta = \int \frac{1}{\mathrm{R}} dx$, the integral being extended to the plastic zone.

396. Since along the plastic zone the moment varies from the elastic limit to the ultimate moment (M_u), it is seen that the basis for the justification of redistribution is the relation between the moment and the curvature which may be written as $\frac{1}{R} = f(M)$.

397. If this relation is known, the maximum possible rotation of the hinge may be evaluated for a given assumed distribution of moments and the assumption will be justified if this maximum is not exceeded. Dr Poologasundram had obtained some interesting results in this direction from tests at Imperial College[32] (Fig. 45).

398. The moment/curvature relation which is assumed to hold at any cross-section of a typical member of a rigid frame is shown diagramatically in Fig. 45, from which it is evident that the curvature ($1/R$) could be expressed as a polynomial in moment M.

Let
$$\frac{1}{R} = A\left(\frac{M}{M_u}\right) + B\left(\frac{M}{M_u}\right)^y \qquad (35)$$

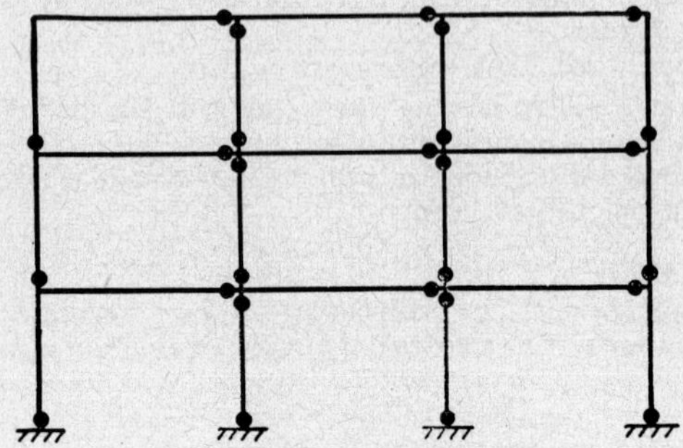

(a) Location of plastic hinges

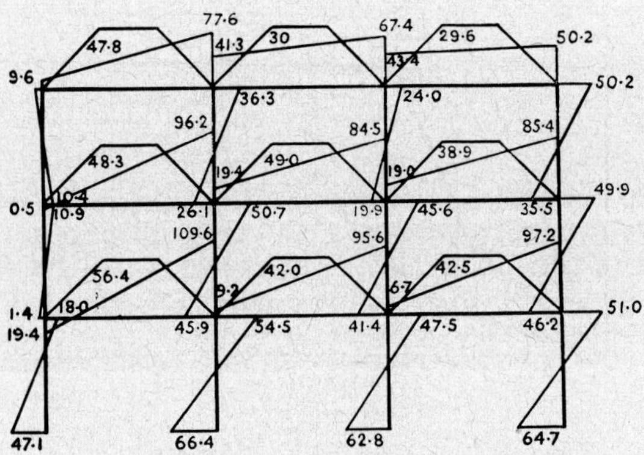

(b) Elastic stage

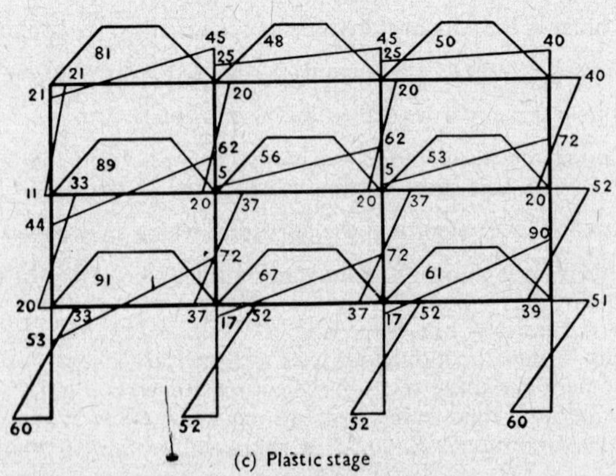

(c) Plastic stage

FIG. 44

where A, B and Y are constants, M_u is the ultimate moment of resistance of the section and M_e is the moment at elastic limit (i.e. corresponding to concrete strain, $e_e = 0{\cdot}001$).

399. $\frac{1}{R} = \frac{e}{nd}$ at all stages within Ob, where e is the maximum compressive strain in the concrete, nd is the depth of the neutral axis. Also $\frac{1}{R} = \frac{M}{EI}$ within the elastic range Oa.

From Fig. 45 $$A = \frac{M_u}{EI},\quad \frac{M_e}{EI} = \frac{e_e}{n_1 d} \quad \text{and} \quad \frac{1}{R_u} = \frac{e_u}{n_u d}$$

where $n_1 d$ and $n_u d$ are the depths of neutral axis at elastic limit and ultimate respectively.

400. substituting for EI, $$A = \frac{M_u}{M_e} \cdot \frac{e_e}{n_1 d} \quad \text{. . . . . (36)}$$

From equation (35), $$1/R_u = A + B$$

Therefore, $$B = 1/R_u - A$$

i.e. $$B = \frac{e_u}{n_u d} - \frac{M_u}{M_e} \cdot \frac{e_e}{n_1 d} \quad \text{. . . . . (37)}$$

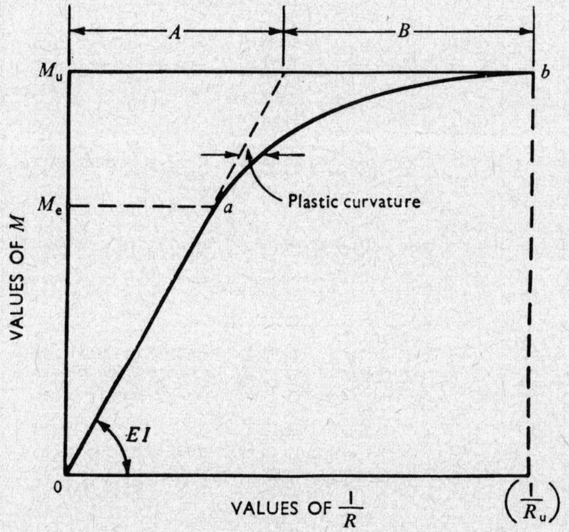

FIG. 45

401. A safe limiting value of $0{\cdot}0035$ is considered satisfactory for e_u and e_e is assumed to be equal to $0{\cdot}001$, if the steel is elastic at this strain, otherwise the concrete strain corresponding to the steel yield strain is used. The value of y is found to depend mainly upon the ratio $\frac{M_e}{M_u}$. The empirical relationship based on the experimental moment/curvature relation has to be obtained for y.

$$y = 3 + 22\left(\frac{M_e}{M_u}\right)^3 \quad \text{. . . . . (4)}$$

402. The justification of the above analysis is seen clearly from the reasonable agreement between measured and computed rotations listed in Tables 16 and 17.

TABLE 16.—COMPARISON OF MEASURED AND COMPUTED VALUES

Beam No.	M_e Kip-in. (1)			M_u Kip-in. (2)			"l_p" in. (3)			θ Radians (4)			Mode of failure
	Computed	Measured	Measured/Computed	Computed	Measured	Measured/Computed	Computed	Measured	Measured/Computed	Computed	Measured	Measured/Computed	
B-1	106·5	107·6	1·01	195·5	193·0	0·99	15·7	13·5	0·86	0·00291	0·00252	0·87	T
B-2	119·5	118·0	0·99	138·2	155·0	1·12	4·7	5·0	1·06	0·00382	0·00396	1·94	T
B-3	81·7	76·3	0·94	87·3	85·0	0·97	1·75	3·5	2·00	0·00342	0·00343	1·00	T
B-4	140·0	147·0	1·05	271·0	310·0	1·14	16·7	15·0	0·90	0·00216	0·00225	1·04	T
B-5	147·5	152·1	1·03	239·0	253·0	1·06	14·0	12·0	0·86	0·00324	0·00358	1·10	T
B-6	105·2	107·2	1·02	197·0	214·0	1·09	16·0	15·0	0·94	0·00181	0·00192	1·06	B
B-7	90·1	93·0	1·02	198·0	232·0	1·17	18·8	14·0	0·75	0·00153	0·00162	1·06	C
B-8	122·1	120·0	0·98	258·4	300·0	1·16	18·2	16·0	0·88	0·00139	0·00156	1·12	C
C-1	106·2	109·0	1·03	194·0	202·0	1·04	10·9	10·5	0·96	0·00198	0·00182	0·92	T
C-2	112·7	112·5	1·00	137·8	148·0	1·07	4·3	4·5	1·04	0·00264	0·00276	1·04	T
C-3	83·4	79·5	0·95	87·1	84·5	0·97	1·75	3·0	1·71	0·00228	0·00260	1·14	T
C-4	144·0	140·0	0·97	274·0	280·0	1·02	11·4	12·0	1·05	0·00169	0·00156	0·92	T
C-5	145·0	149·0	1·03	239·0	248·0	1·04	9·95	9·0	0·90	0·00226	0·00259	1·15	T
C-6	106·0	109·0	1·03	201·0	218·0	1·08	10·2	11·0	1·08	0·00127	0·00142	1·04	B
C-7	91·0	99·0	1·09	198·7	243·0	1·22	13·0	12·0	0·92	0·00118	0·00145	1·23	C
C-8	123·1	118·0	0·96	264·0	310·0	1·17	12·8	13·0	1·01	0·00102	0·00125	1·22	C

TABLE 17.—COMPARISON OF MEASURED AND COMPUTED VALUES

Beam or Strut No.	M_e Kip-in. (1)			M_u Kip-in. (2)			"l_p" in. (3)			θ Radians (4)			Mode of failure
	Computed	Measured	Measured/Computed	Computed	Measured	Measured/Computed	Computed	Measured	Measured/Computed	Computed	Measured	Measured/Computed	
T-1	84·4	88·5	1·05	113·0	138·0	1·22	8·80	8·0	0·91	0·00384	0·00442	1·15	T
T-2	100·3	112·0	1·10	191·0	215·0	1·12	16·4	14·0	0·86	0·00329	0·00324	0·99	T
T-3	119·6	116·0	0·97	266·0	290·0	1·09	19·2	16·0	0·84	0·00184	0·00222	1·20	I
T-4	124·5	132·0	1·06	269·0	300·0	1·11	18·4	15·0	0·82	0·00124	0·00162	1·30	I
S-1	108·8	103·8	0·95	157·5	173·4	1·10	11·2	12·0	1·07	0·00385	0·00365	0·95	T
S-2	102·7	95·0	0·93	196·1	205·5	1·05	17·3	19·0	1·10	0·00311	0·00295	0·94	T
S-3	94·2	87·0	0·92	195·0	217·5	1·12	18·6	20·0	1·08	0·00250	0·00325	1·30	Near Balanced
S-4	80·5	75·5	0·94	170·5	188·9	1·11	19·0	22·0	1·16	0·00255	0·00300	1·17	C
S-5	46·0	43·0	0·94	138·0	148·0	1·07	24·0	26·0	1·08	0·00310	0·00325	1·05	C
S-6	10·0	—	—	116·0	128·0	1·10	32·9	36·0	1·09	0·00350	0·00395	1·03	C

Dr K. A. Everard (Head, Department of Civil Engineering, University of the West Indies, St Augustine, Trinidad) wrote that the Report recorded the tremendous progress which had been made during the last ten years in the application of ultimate load theory to the design of reinforced concrete structures. The theory had now become a powerful design tool.

403. At the University of the West Indies they were determining the distribution of stress under flexure in reinforced concrete beams and they were calculating, from first principles, the moment/curvature relationships.

404. The method for the determination of stress distribution was an indirect one where specimens were tested at uniform rates of strain.[99] The stress at a point, in a concrete beam undergoing uniform increase of curvature, depended on the values of the strain and rate of strain at that point.

405. If it were assumed, however, in a reinforced concrete section that a particular stress might be identified with a particular strain and that the stress/strain relationship might be adequately specified by a polynomial in n terms, then together with the Bernoulli-Navier assumption of plane sections a theory of bending of the following form might be obtained

$$M = \sum_{r=1}^{n} Ar\left(\frac{d^2y}{dx^2}\right)^r \quad . \; . \; . \; . \; . \; . \; . \; . \; . \; . \quad (37)$$

where M = moment of resistance

and Ar is a function of the neutral axis depth.

406. By adopting a simple parabolic distribution, expressions might be obtained in a dimensionless form to describe the moment/curvature relationships for rectangular beams.[100]

407. Thus for a beam of breadth b and effective depth d, we obtain

$$\alpha = \frac{E_0}{\mu}\left(\frac{d}{R}\right) = \frac{12}{n^3}\left[\frac{n^2}{2}-(1-n)mp\right] \quad . \; . \; . \; . \; . \; . \quad (38)$$

$$M(\alpha) = \frac{1}{\mu}\left(\frac{M}{bd^2}\right) = \left[\frac{n^3}{3}+(1-n)^2mp\right]\alpha-\frac{n^4}{16}\alpha^2 \quad . \; . \; . \; . \quad (39)$$

which apply up to the yield point of steel when:

$$\alpha=\frac{3}{n}\left[1-\sqrt{-\frac{8n_{\text{ult}}}{9n}}\right] \quad . \; . \; . \; . \; . \; . \; . \; . \quad (40)$$

and

$$M(\alpha) = \frac{n^3}{3}\cdot\alpha\cdot+mp\cdot N\cdot(1-n)-\frac{n^4}{16}\alpha^2 \quad . \; . \; . \; . \; . \; . \quad (41)$$

408. Expressions (37) and (38) were continuous and applied throughout the range of moment up to the ultimate where no yield of steel took place. In an under-reinforced section they applied up to the yield of steel, whereupon a discontinuity occurred and (40) and (41) were applicable to failure.

409. It could be shown that the departure from linearity of the moment/curvature relationship due to non-linear stress distribution (time independent) was not great when there was no yield of tensile reinforcement. With reducing proportions of reinforcement, however, the departure from linearity increased, the effective "ductility" of a section having an inverse relationship to the quantity of steel.

410. It was possible to calculate the angle of discontinuity (corresponding to rotation of a plastic hinge) which occurred under various bending moment distributions. For an over-reinforced section this was not great but for members with small percentages of steel it could be quite appreciable.

411. If the moment were assumed to remain appreciably constant over a length lp and at the ultimate value, the simplest case occurred and for $lp=d$

$$\phi = \left(\frac{E_0}{\mu}\right)\theta p - \left\{\alpha_{\max}-\frac{M(\alpha)}{B(n)}\right\}$$

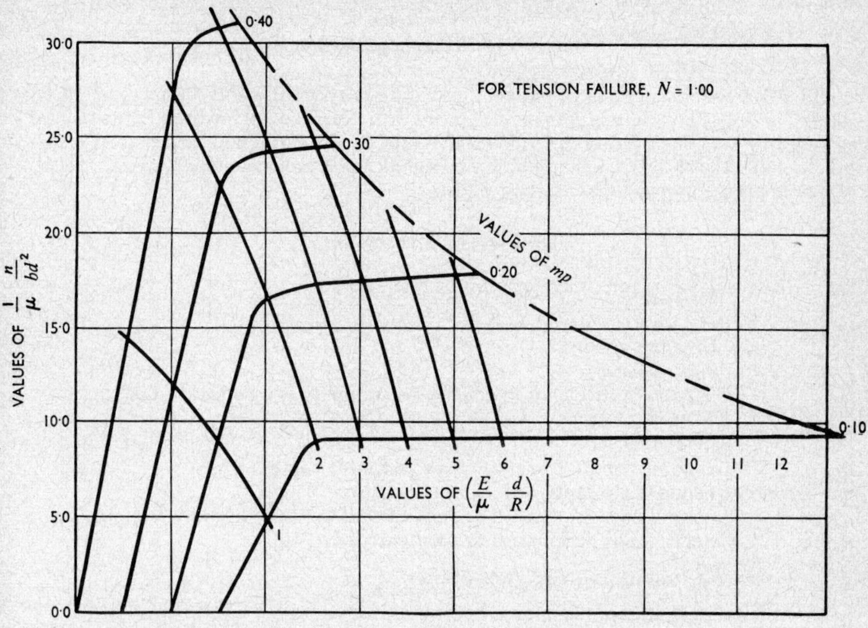

FIG. 46.—MOMENT/CURVATURE RELATION FOR UNDER-REINFORCED BEAMS

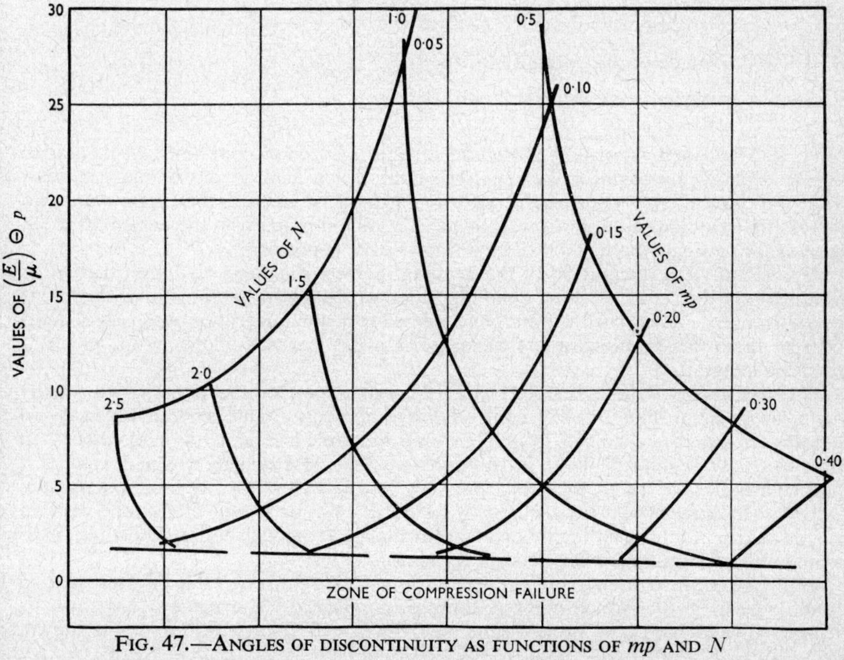

FIG. 47.—ANGLES OF DISCONTINUITY AS FUNCTIONS OF mp AND N

Where

$$B(n) = \left[\frac{n_0^3}{3} + (1-n_0)^2 mp\right]$$

412. The graphical forms of moment/curvature relationships and ϕ were reproduced in Figs 46 and 47 and were in reasonable accord with experimental forms. Experimental strains, however, included both strains which might represent a non-linear stress distribution and creep strains which might not be significant in redistributing stresses. Work was now proceeding to separate these effects.

Dr Everard's notation

E_0 $= \left(\frac{df}{de}\right)_{e=0}$ for concrete
m $= E_{steel/E0}$
e = concrete strain
f = concrete stress
l_p = length over which appreciable discontinuity is assumed to take place
$M(\alpha)$ $= \frac{I}{\mu}\frac{M}{bd^2}$
n = ratio of depth of neutral axis to effective depth d
n_0 = neutral axis depth ratio for elastic case
n_y = neutral axis depth ratio at yield of steel
n_{ult} = neutral axis depth ratio at ultimate load
N $= \frac{ty}{m\mu}$ a dimensionless stress ratio
ty = yield stress of reinforcement
α $= \frac{E_0}{\mu}\frac{d}{R}$
θ_p = angle of discontinuity
$\phi = \left(\frac{E_0}{\mu}\right)\theta_p$ = angle of discontinuity factor
μ = Cylinder crushing strength

Mr L. Wadsworth (Assistant Structural Engineer, Luton Corporation) wrote that to him as a practising engineer the simplified limit design method was of most interest. The proposals therein appeared straightforward but it seemed that there were two factors that would prejudice its acceptance as common practice in the design office, at least as far as single-storey factory type frames were concerned.

414. Firstly the integration of the bending moment diagrams to obtain the values of hinge rotations could become quite laborious, especially for asymmetrical single-bay portal frames. When such a frame had pinned feet, the integration was very tedious because there was no point where the slope of a member was known, from which to start the integration.

415. Secondly, with a flexible frame, the summation of the terms $\bar{X}\delta ik$ usually entailed the small differences of relatively large numbers. This made it necessary to use greater accuracy than was available with a slide rule in calculating δik, δo etc.—to the 5th or 6th significant figure in some cases. Thus in Example 3, which was not a particularly flexible frame, the coefficients δik were calculated to 5 decimal places.

416. He thought that the method of coupling a plastic design of sections with an elastic distribution of bending moments (utilizing a 15% redistribution where required) would remain popular for this class of frame.

417. Again, in Example 3 of the Report, on working through this example he had found several points which showed some inconsistency. The value $\delta 1b3b$, and its reciprocal value $\delta 3b1b$, in Table 8 came to $0{\cdot}02414 \times 10^{-6}$ not $0{\cdot}0225 \times 10^{-6}$ as shown.

Incidentally, the arrows indicating the direction of action of the applied moments in condition M_3 of Fig. 21 appeared to go in the wrong direction.

418. The values in Tables 8 and 9 for δik and δo seemed to represent the rotations at both hinges b_1, both hinges b_2 etc., so that the final figures for actual rotations gave twice the rotation at any one hinge. These figures were then compared with the θp value calculated for hinge $1b$, apparently being the rotation at one hinge $1b$ only.

References

34. H. M. Mackay, P. Gillespie, and C. Lelnau, "Strength tests on concrete-encased steel beams", Engg News Record, 3 May and 28 June, 1923. Also Conc. and Construct. Engg, vol. 19, pp. 5–8, 1924.
35. I. M. Viest, C. P. Seiss, J. H. Appleton, and N. M. Newmark, "Channel shear connectors and composite tee-beams", Univ. Illinois, Engg Exp. Sta. Bull. 405, 1952. Other tests in Proc. Amer. Conc. Inst., vol. 57, 1960, pp. 609–628.

35a. "Composite construction for steel-framed buildings", Brit. Construct. Steelwork Assoc., Publ. No. 15, 1961 and No. 17, 1962.

36. H. L. Su, Reply to discussion on "Statistical approach to structural design", Proc. Instn civ. Engrs, vol. 15, March 1960, pp. 299–301.
37. H. L. Su, "Some problems in the philosophy of structural design", Proc. Instn civ. Engrs, vol. 23, Aug. 1962, pp. 391–404.
38. H. L. Su, Contribution to the Discussion of "Use of plastic theory in the design of a fabricating shop", Prof. Sir J. F. Baker, Prof. M. R. Horne, and L. J. Johnson. To be published in Proc. Instn civ. Engrs.
39. H. L. Su, "Design by quantitative safety factor", Trans. ASME, vol. 82, B, Nov. 1960.
40. H. L. Su, "Why plastic design" Lecture Note, to be published.
41. A. L. L. Baker, "A plastic design theory for ordinary reinforced and prestressed concrete shell roofs", Mag. Conc. Res., Jul. 1950, pp. 27–34.
42. A. L. L. Baker, "A plastic theory of design for ordinary reinforced and prestressed concrete including moment redistribution in continuous members", Mag. Conc. Res., Jun. 1949, pp. 57–66.
43. A. L. L. Baker, Discussion and reply, "Recent research in reinforced concrete and its application to design", J. Instn civ. Engrs, vol. 35, Feb. 1951, pp. 298–329.
44. S. R. Sparkes, J. C. Chapman, and A. C. Cassell, "Structural experiments on a building of composite design", 6th Cong. Int. Assn Bridge Struct. Engg, Stockholm, 1960, pp. 479–492.
45. H. L. Su, "Philosophical aspects of structural design", Proc. ASCE, J. Struct. Divsn, June 1961, pp. 1–16.
46. D. E. Drucker, W. Prager, and J. Greenberg, "Extended limit design theorems for continuous media", Q. appl. Math., vol. 9, 1952, pp. 381–389.
47. H. J. Greenberg and W. Prager, "On limit design of beams and frames", Trans. ASCE, vol. 117, 1952, pp. 447–458.
48. W. Prager, "Limit analysis and design", Proc. Amer. Conc. Inst., vol. 50, 1954, pp. 297–304.
49. W. Prager and P. C. Hodge, "Theory of perfectly plastic solids", Wiley, 1951.
50. B. G. Neal, "The plastic methods of structural analysis", London, 1956.
51. J. F. Baker, M. R. Horne, and J. Heyman, "The steel skeleton", vol. 2, Cambridge Univ. Press, 1956.
52. H. L. Su, Discussion on "Commentary on plastic design in steel", Proc. ASCE, J. Engg Mech. Divsn, Jan. 1960, pp. 167–168.
53. (Deleted).
54. C. W. Yu, M. Corbin, and E. Hognestad, "Reinforced concrete design in the U.S.S.R.", J. Amer. Conc. Inst., vol. 31, No. 1, July, 1959, pp. 65–69.

55. L. L. Jones, "Ultimate load analysis of reinforced and prestressed concrete structures", Chatto and Windus, London, 1962, p. 246.

56. Norms and rules in construction (S.N.i.P.) (in Russian) Part II, Gosstrojizdat, Moscow 1955.

57. Proposed revision of building code requirements for reinforced concrete (ACI 318–56). J. Amer. Conc. Inst., vol. 59, No. 2, Feb. 1962, pp. 145–276.

58. T. Y. Lin, "Load factors for prestressed concrete bridges", Proc. ASCE, vol. 83, J. Struct. Divsn, July 1957, pp. 1315-1–1315-18.

59. Y. Guyon, "Dimensionnement à rupture d'une ossature à cadres en béton armée", CEB Bulletin No. 30/1961, pp. 21–68.

60. G. Macchi, "Proposition de calcul basée sur la théorie des rotations imposées", CEB Bulletin No. 21/1960, pp. 121–158.

60a. G. Macchi, "Etude expérimentale des poutres continues précontraintes dans le domaine plastique et à la rupture", 2nd Cong. Fed. int. Précontr., Amsterdam, 1955, pp. 501–543.

61. A. L. L. Baker, "A general equation for frame analysis", Conc. and Construct. Engg, vol. 56, No. 3, March 1961, pp. 97–103.

62. V. Petcu, "The optimum redistribution principle", Indian Conc. Jnl, vol. 35, No. 7, July 1961, pp. 259–260 (see also vol. 36, No. 1, Jan. 1962).

63. M. Z. Cohn, "Limit design of reinforced concrete structures for maximum yield safety". (To be published in the Indian Concrete Journal).

64. M. Z. Cohn, "On the redistribution factors of simple beams and frames". 3rd Scientific Conference of the Building Institute, Bucharest, 1957.

65. V. Petcu, "Plastic analysis of reinforced concrete continuous beams" (in Rumanian) Studii şi Cercetări de Mecanică Aplicată (Bucharest), vol. 12, No. 6, 1961, pp. 1321–1334.

66. Beton-Kalender, W. Ernst und Sohn, Berlin 1956 (in German).

67. M. Z. Cohn, "On a simple rule for the limit design of reinforced concrete continuous beams". (To be published.)

68. Provisions for the analysis of redundant reinforced concrete structures allowing for the redistribution of stresses. (In Russian) N.I.I.Zh.B.-A.S.i.A., U.S.S.R., Gosstrojizdat, Moscow, 1960.

69. M. D. Hangan, "Order of occurrence and rotation of plastic hinges in structures acted simultaneously by permanent and proportionally increasing loads", Revue de Mécanique Appliquée (Bucharest), vol. 6, No. 2, 1961, pp. 229–240.

70. M. Z. Cohn and V. A. Petcu, "An experimental investigation on the carrying capacity of fixed-ended reinforced concrete beams" (in Rumanian). Bulletin of the Building Research Institute (Bucharest), vol. 2, No. 3, 1961, pp. 71–84.

71. M. Z. Cohn and V. A. Petcu, "On the influence of steel percentage on the plastic adaptability of redundant reinforced concrete beams", Bull. Acad. Polo. SCI., Warsaw. Série des Sciences Techniques, vol. 8, No. 11–12, 1960, pp. 713–721.

72. V. A. Petcu and M. Z. Cohn, "On the moment redistribution in redundant highly-reinforced concrete beams", Bull. Acad. Polo. SCI., Warsaw. Série des Sciences Techniques, vol. 9, No. 10, 1961, pp. 593–600.

73. A. H. Mattock, "Redistribution of design bending moments in reinforced concrete continuous beams", Proc. Instn civ. Engrs, vol. 13, May 1959, pp. 35–46.

74. Инструкция по расчету статически неопределимых железобетонных конструкций с учетом перераспределений усилнй.
(Instruction for the design of statically indeterminate reinforced and prestressed concrete structures, 2nd edition, Academy of Building and Architecture of the USSR Research Institute for Plain and Reinforced Concrete, Moscow, 1961).

75. (Deleted).

76. R. H. Wood, "A preliminary study of composite action in framed buildings", Int. Assn Bridge Structural Engg, vol. 15, pp. 247–265.
77. R. H. Wood, "Plastic and elastic design of slabs and plates", Thames & Hudson, 1961.
78. K. Terzaghi and R. B. Peck, "Soil mechanics in engineering practice", Wiley, New York, 1948, pp. 422 and 433.
79. "Building Research, 1958", Report of the D.S.I.R. Building Research Station for the year 1958, H.M.S.O., 1959.
80. "Report on structural safety", J. Instn Struct. Engrs, 1955, vol. 33, p. 141.
81. A. L. L. Baker, "Ultimate load design of reinforced and prestressed concrete frames", Proc. Symp. Str. Concr. Struct., London, May 1956, pp. 277–304, C. & C.A., London, 1958.
82. G. Macchi, Discussion to Session I, Proc. 3rd Cong. Féd. int. Précontr., Berlin, 1958, pp. 16–21, C. & C.A., London, 1959.
83. E. W. Bennett, "The redistribution of bending moment in continuous prestressed concrete beams", Ph.D. Thesis, University of Leeds, 1960.
84. P. B. Morice and H. E. Lewis, "The ultimate strength of two-span continuous prestressed concrete beams as affected by tendon transformation and untensioned steel", Proc. 2nd Cong. Féd. int. Précontr., Amsterdam, 1955, pp. 561–586, C. & C.A., London, 1958.
85. T. Y. Lin, "Strength of continuous prestressed concrete beams under static and repeated loads", Proc. Amer. Concr. Inst., vol. 51, 1955, pp. 1037–1053.
86. (Deleted).
87. P. B. Morice, "Linear structural analysis", Thames and Hudson, London, 1959.
88. E. Lightfoot, "Moment distribution", Spon, London, 1961.
89. (Deleted).
90. R. H. Wood, Contribution to the Discussion, Ref. 77.
91. A. L. L. Baker, "The ultimate load theory applied to the design of reinforced and prestressed concrete frames", Concrete Publications Ltd, p. 78.
92. R. H. Wood, "The stability of tall buildings", Proc. Instn civ. Engrs, vol. 11, Sept. 1958, pp. 69–102.
93. E. Freyssinet, "The birth of prestressing", C. & C.A. Library Translation No. 59, pp. 12–15.
94. L. J. Mensch, Discussion of Final Report of the Special Committee on Concrete and Reinforced Concrete, Paper 1424, Trans. ASCE, Dec. 1918.
95. Riccardo Morandi, "Engineering and Architecture", J. roy. Soc. Arts, vol. 110, 1962, pp. 75–88.
96. G. Brock, Discussion of Title 55–66, J. Amer. Conc. Inst., Dec. 1959, vol. 55, pp. 1571–1575.
97. A. J. Ockleston, "Loading tests on reinforced concrete slabs spanning in two directions", Portland Cement Instn (S.A.), 1958.
98. C. Benito, "Essais comparatifs pour le choix d'une forme résistante", Symp. Shell Research, RILEM-IASS, Delft, 1961, pp. 262–266.
99. K. A. Everard, Discussion of "Further research in reinforced concrete and its application to ultimate load design", by A. L. L. Baker, Proc. Instn civ. Engrs, vol. 3, Part 3, April 1954, pp. 309–311.
100. K. A. Everard, "The flexural rigidity of reinforced concrete". (To be published).

Reply to written discussion

The Committee greatly appreciates the wide interest which has been taken in the Report, and the constructive contributions to the discussion which have been received. These, together with the Report, should provide research workers and designers with a fairly comprehensive statement of our present knowledge of the subject.

420. Opinions of the contributors range from those who wish to apply plastic hinge theory developed for structural steelwork to reinforced concrete, to those who still prefer to use elastic analysis. Those familiar with Russian and Danish practice accord most closely to the approach recommended in the Report, which recognizes the special structural characteristics of reinforced concrete.

421. Several contributors object to the procedure of assuming hinge positions and checking rotations to ensure moment-deformation compatibility as too complicated. It is necessary to have a sound basic general theory. Simple design rules will follow from it eventually and much design be done by their use, leaving only special structures to be analysed from first principles by those competent to do so. For instance, in the case of continuous beams:

Notation

M = bending moment
l = span of beam
EI = flexural rigidity
b = breadth of beam
d = effective depth of beam
t_u = design ultimate stress of steel
c_u = cube strength of concrete
ϕ = diameter of bars
k_r = an empirical constant
$p = \dfrac{\text{area of tensile reinforcement}}{bd}$

422. The following is intended to provide an example of the type of simple design rule which may be established from plastic hinge or ultimate load theory. As more research results become available and design experience grows, more such rules can be produced. As the distribution of bending moment can more easily and accurately be defined at the ultimate stage of load in terms of safe limiting values of parameters, eventually design theory should become simpler and at the same time more accurate, satisfactory serviceability at working load being ensured at the same time. Economy can also be made since, for the various cases of loading, the plastic hinges reduce the more extreme values of bending moment as obtained by the elastic theory. The practical engineer, in comparing the existing standard formulae derived from elastic theory with ultimate load theory, should have in mind such simplifications of basic ultimate load theory as this.

423. Consider a continuous beam of any number of spans greater than three having the following characteristics:

(*a*) The flexural rigidity *EI* after cracking is approximately constant.

(*b*) The live load and dead load are uniformly distributed and the live load does not exceed 50 per cent of the dead load.

(*c*) No span is more than 15 per cent longer than the shortest span. The penultimate span may be 15 per cent longer than the end span, but not equal to it, or smaller.

424. All support sections are designed to have an ultimate bending moment value of $\frac{M}{2}$ at L_2 (§ 53) where M is the value of the maximum "free" bending moment in either of the adjacent spans. Mid-span sections are designed for moments L_1 at ultimate load.

425. Fig. 48 shows the various cases of load for a four-span beam having the following characteristics:

Case 1: Maximum elastic moment and inelastic rotation at support 2.
Case 2: Maximum elastic moment at mid-span 1–2.
Case 3: Maximum elastic moment and inelastic rotation at support 1. This has the following conditions which are the worst for all supports for all possible cases of loading, including a beam of any number of spans greater than three.

(*a*) Hinge Rotation at support 1 at ultimate load $= \frac{\cdot 19Ml}{EI}$ approximately.

(*b*) Steel stress at working load at support 1, which is assumed to be half ultimate load $= \frac{\text{yield stress.}}{1{\cdot}4}$

(*c*) The ultimate bending restraint at support 1, when designed to be $M/2$ is 30 per cent less than the value obtained by elastic design.

426. By using safe values of parameters (§ 53) and equating the maximum hinge rotation to the permissible rotation, it is found that the permissible rotation will not be exceeded, provided

$$\frac{l}{d} \ngtr \frac{1}{k_r}\left(\frac{c_u}{p.t_u}\right)^2$$

427. A safe value of k_r deduced from existing test results is 10. As more research is done, it may be found that some favourable modification can be recommended.

428. It is also found that, by using the C.E.B. crack width formulae[105], cracks at support section 1 in Case 3 do not exceed $\frac{8}{1000}$ in., i.e. 0·2 mm at working load, provided

$$\frac{\phi.t_u}{p}\left(\frac{c_u}{6000}\right) \ngtr 5{\cdot}0\times 10^6$$

$\left(\text{assumed value of } k \text{ [ref. 2]} = 3\times 10^{-3}\text{, i.e. greatest recommended value, } c_t \text{ assumed} = \frac{c_u}{10}\right)$

Design rule

429. A design rule may therefore be established for any continuous beam having the following characteristics:

(*a*) The flexural rigidity *EI* after cracking is approximately constant.

(*b*) The live and dead loads are uniformly distributed and the live load does not exceed 50 per cent of the dead load.

(*c*) There are at least four spans. No span is 15 per cent longer than the shortest span. The penultimate span may be 15 per cent longer than the end span, but not equal to it or shorter.

The minimum moment value at the support may be assumed to be $M/2$, where M is the maximum value of the free bending moment in either of the adjacent spans, provided

$$\frac{l}{d} \ngtr \frac{1}{10}\left(\frac{c_u}{p \, . \, t_u}\right)^2$$

430. Permissible values of ultimate restraint moments are $\ngtr L_1$ (§ 53) at mid-span sections and $\ngtr L_2$ (§ 53) at support sections. Also, crack width at working load cannot exceed 0·2 mm, provided

$$\frac{\phi \, . \, t_u}{p}\left(\frac{c_u}{6000}\right) \ngtr 5{\cdot}0 \times 10^6$$

(ϕ inches, t_u and c_u lb/sq. in.)

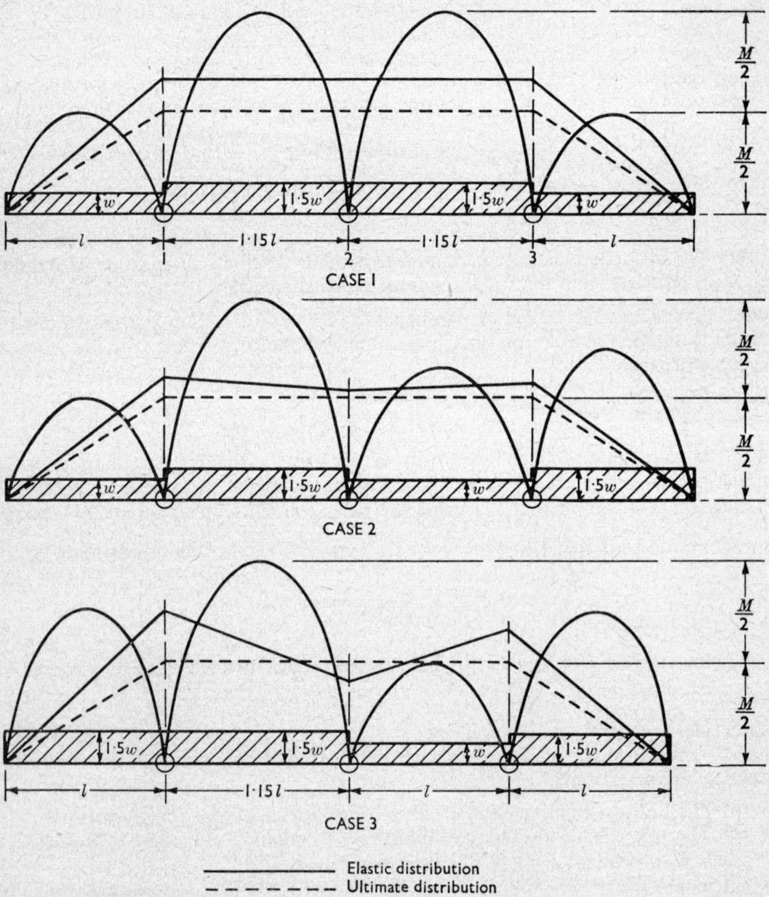

Fig. 48

431. If the ultimate load is assumed to be twice the working load, the steel stress at working load at support 1 is $< \frac{f_u}{1 \cdot 3}$ for the worst possible distribution of live load.

432. Mr M. Z. Cohn and Mr Milík Tichý, in their contributions, have made valuable proposals for the practical application of ultimate theory. It is interesting to note that the latest edition of the Soviet Instructions for Reinforced Concrete Design recommend the use of equation (5), which was first published in 1951[104].

433. Mr E. W. Bennett's comments show that in prestressed beams, as might be expected, because of the steeply rising moment-rotation curve between L_1 and L_2, actual ultimate bending strength of continuous beams can be 10 per cent greater than the value calculated by the simplified Limit Method, when moments are assumed equal at L_1 and L_2. There is no reason why the Rigorous Method B should not be used together with an established moment-rotation law, as proposed. Such a method would, however, be complicated if applied to frames. The simplified Limit Method gives safe results without the ultimate strength being too greatly under-estimated, if L_1 and L_2 are related, as in Fig. 10. Optimistic conclusions in regard to full redistribution should not be made from a small number of tests. There is probably a scatter of results yet to be revealed. The example shown in Fig. 48 indicates the influence of slenderness and of the distribution of live load on results. The worst cases for rotation should be tested [102].

434. Mr L. L. Jones asks for the reasons for not applying the Cambridge Steel Skeleton methods to reinforced concrete. The latter material is too brittle and too variable in deformation characteristics to assume that full redistribution of bending moments can always take place. (Refer to example in Fig. 48. Tests should be made for cases in which all parameters influencing rotations have the worst value.) Moreover, as Mr R. H. Wood has shown, there is a risk of frame instability. Only by specially locating the positions of plastic hinges at the statically determinate stage, can a safe design for stability be worked out. The Report suggests designing a particular system of hinges and elastic corner restraints to avoid this danger. A similar procedure might be followed with advantage in designing structural steelwork.

435. The difficulty of a free choice of Load Factor Value is admitted and eventually perhaps structures will have to be classified.

436. Mr R. E. Rowe has indicated various research subjects requiring attention. His proposals are strongly supported. Deformation characteristics of all forms of restraints in all kinds of frame members, including joints of pre-cast members, are required.

437. The Load Factor proposals, criticized by Professor M. R. Horne, should perhaps be the subject of a special report on security. To comment very briefly: Reinforced Concrete design practice has, over several decades, produced thousands of structures which have proved to be safe when reinforced with steel stressed at working load to half the yield stress or even a little more. This is so, even in such structures as simply supported gallery beams frequently called upon to support full live load, with no reserve of strength, by bending moment redistribution or other aids not allowed for in design calculations. A Load Factor of 2·0 therefore appears to have been established as adequate, provided the conditions of § 68 are applied, and the designer takes a responsible attitude in his assessment of maximum load. It is a sound principle to assume statistically established minimum material strengths and maximum loads in calculations, in conjunction with a low Load Factor value, such as 2. There is then less risk of encroachment by contractors on the Load Factor Value which is known to be small by tolerating poor quality workmanship or permitting over-loading.

438. Professor Horne's diagram, Fig. 41, helps to clarify the Report's recommendations in regard to instability. The Limit Design assumptions and procedure may be summarized, as follows:

(*a*) When ultimate load is applied to the idealized frame, sufficient plastic hinges form to make it statically determinate.

(*b*) If the assumed positions of the hinges satisfy the compatibility equations, the hinges will finally form, at the statically indeterminate stage, at the assumed sections as load is increased at uniform intensity throughout the frame[103].

(*c*) A frame having real moment deformation characteristics will be stronger than each case of idealized frame designed for each possible case of loading and the Limit assumptions L_1 and L_2.

(*d*) To avoid frame instability, the idealized frame will be so designed that plastic hinges at the statically determinate stage of load isolate each storey. Each storey can then be regarded as an independent frame receiving known loads, support and bending moments from adjacent storeys.

(*e*) Each independent storey, being statically determinate, can be designed for stability by providing that at least one column-to-beam joint remains elastic in the statically determinate stage. The problem of over-all frame instability is thus avoided.

(*f*) When columns or beams are slender, the sway of each separate storey must be calculated and the additional bending in each column, due to an assumed eccentricity of axial loads, allowed for. It must be ascertained that the calculated sway and deflexion of each column due to horizontal forces unsymmetrical and eccentric vertical load is less than the assumed value.

439. It may well be that in building frames (ignoring all sway resistance due to partitions), the vertical load bending moments due to eccentricity caused by sway and buckling, is sometimes greater than has been realized, so that Mr R. H. Wood has given a timely warning. However, the procedure in (*d*) and (*f*) above will ensure a design which is stable when the frame has sufficient hinges to be statically determinate. Moreover, the hinges in the idealized frame will form as assumed, if the compatibility equations are satisfied. The actual frame may behave differently from the idealized frame, but will be stronger, unless some abnormal condition of brittleness exists.

440. Mr H. L. Su's philosophy of design appears to be a little ostrich-like. Surely it is wise to reflect on the state of failure in order to avoid it. § 72 points out the importance of design for serviceability. A slender column or a highly prestressed and reinforced beam giving no signs of failure at working load, and which is not over-stressed, may collapse without warning at slight overload, since there is a non linear relationship between load and stress.

441. Mr R. H. Wood has given a warning in regard to the danger of instability due to sway. The Report, in its recommendations for designing for instability, takes account of both buckling and side sway (see § 76). It is safer to investigate deformation and bending compatibility and to base calculations on such a trial analysis than to increase the Load Factor by an arbitrary amount which may or may not provide the additional strength and stiffness required.

442. The Committee certainly agrees with Dr Gartner in regard to water-retaining structures. They should be designed for serviceability at low stresses with no risk of cracking.

443. The Committee, in the course of its work, did not give any consideration to composite construction, as discussed by Mr A. N. Proctor. The inter-action of steel and concrete through various types of shear connector presents problems of a special nature. Mr Proctor's contribution is welcome, but a reply would be better given by a committee appointed to study research in this subject. In frames, the continuity at joints and characteristics of plastic hinges would certainly require special empirical treatment.

444. Mr R. J. Dailey asks for information on the rate of increase of θ_p with age. Under normal load, the distribution of bending moments in a structure may change due to differential creep rates at various critical sections. When ultimate load acts, much greater deformations take place, as 80 to 100 per cent of ultimate load is applied, so that creep strains due to normal load have little influence on the ultimate distribution of bending moments and accompanying hinge rotations. Long-term stress/strain

tests on concrete indicate that rotations due to slow relative settlement of columns could safely be assumed to be greater than short-term rotations due to over-load.

445. The permissible rotation for ultimate load must be reduced by the settlement, but could be based on an ultimate strain value suitable for slowly applied load. Concrete, over-stressed to 75 per cent of ultimate strength for a long period, may be weakened[5].

446. § 68(*d*) refers to special design requirements when fatigue is expected. § 66(*f*) refers to risk of weakening of the structure by any cause, not only fatigue.

447. The value of n_u is based on $C=T$. It is assumed that the steel yields until the concrete crushes.

$$e_{su} = 0{\cdot}0035 \times \frac{0{\cdot}44}{0{\cdot}56} = 0{\cdot}00275$$

$$e_s \text{ at } L_1 = \frac{80{,}000}{30 \times 10^6} = 0{\cdot}0026$$

448. The steel therefore extends inelastically $0{\cdot}00275 - 0{\cdot}0026 = 0{\cdot}00015$ per unit length when the concrete crushes. Equating $C=T$, assuming a parabolic distribution of concrete stress, is a simple way of calculating n_u.

449. Dr G. Brock proposes that it should be assumed that the fully plastic collapse load of a frame could be reached by precautions in detailing. In a structural steel frame "fully plastic collapse load" means that every critical section develops its full plastic moment before collapse of the frame. In a reinforced concrete frame there is no restriction to sections of uniform standard bending strength. Reinforcement can easily be adjusted locally to give the required local strength and stiffness. The distribution of bending movements can be varied at will to produce economy of design, both in regard to the quantity of reinforcement used and the ease with which construction can be carried out. For this purpose, adjustments to elastic distributions of moments are already used to advantage. The Report gives the basic theory from which such adjustment rules may be made. Before it is assumed that collapse loads for reinforced concrete frames can be calculated by simple plastic theory which assumes adequate ductility for full rotation, a very large number of tests must be carried out in which maximum hinge rotations are required in sections of minimum permissible rotation, including columns subject to compression and bending, as well as beams.

450. In regard to Mr W. J. R. Smyth's comment, it may be that re-distribution of bending moments to required limits can be obtained in normal structures without excessive cracking or rotations. But this must be proved by significant test results that verify a theory which may be applied generally to a well defined range of structures likely to occur in practice.

Acknowledgements

K. Poologasundram, Ph.D. and K. A. Everard, Ph.D., are thanked for useful contributions.

References

101. European Committee for Concrete. Practical recommendations for use by designers.
102. A. L. L. Baker, "A general equation for frame analysis". Concr. constr. Engng, vol. 56, March 1961, pp. 96–103.
103. A. L. L. Baker, "Ultimate load theory for concrete frame analysis". Proc. Amer. Soc. civ. Engrs, vol. 85, J. Struct. Divn, November 1959, pp. 1–29.
104. A. L. L. Baker, "Recent research in reinforced concrete, and its application to design". J. Instn civ. Engrs, vol. 35, February 1951, pp. 262–298.
105. A. L. L. Baker, "European Committee on Concrete". Report on a meeting at Vienna in April 1959. Reinf. Concr. Rev., vol. 5, no. 4, December 1959, pp. 247–264.

Supplementary informal discussion

On Tuesday, 13 November, 1962, an informal discussion on the Research Committee's Report on 'Ultimate load design of concrete structures' was introduced by Professor A. L. L. Baker, D.Sc.(Eng.), M.I.C.E. (Professor of Concrete Structures and Technology at the Imperial College of Science and Technology), chairman of the Committee which prepared the Report. The following summary of the introductory remarks, the comments made by the several speakers in the discussion, and Professor Baker's reply, has been prepared by Mr Chas. E. Reynolds, B.Sc.(Eng.), M.I.C.E. (Managing Editor, Concrete Publications Ltd).

Chairman's introduction

Mr R. E. Rowe presided and, in opening the meeting, explained that the Report prepared by the Research Committee of the Institution on the 'Ultimate load design of concrete structures' was published in the Proceedings for February 1962 (vol. 21, pp. 400–442) and a written discussion thereon had been published in the Proceedings for October 1962 (vol. 23, pp. 225–269). The Research Committee's reply was published in the Proceedings for November 1962 (vol. 23, pp. 532–537). The purpose of the present meeting was to continue the discussion informally.

Professor Baker's introductory remarks

Professor Baker in his introductory remarks said that the Research Committee had reported on the present position regarding the theory, practice and research of limit design of reinforced concrete structures. A warning was given in the Report to the effect that the engineer should study the behaviour of a structure at all stages of loading up to failure, but there was no single absolute rule of approach for the structural designer to follow. Sometimes, as in the case of tanks, the most important criterion was to limit the tensile stress. In a building frame, an adequate margin of safety against instability might be the vital consideration. For some structures a rigorous analysis might be necessary, but for others an approximate calculation might be satisfactory. It was for the supervising designer to judge the degree of computational accuracy necessary in a given case, and he must be aware of the possible variations and significance of the assumptions made. Although the Report dealt mainly with ultimate conditions as the criterion for design, satisfactory serviceability must also be given attention.

454. It was convenient to express the basic requirements of frame analysis by a compatibility equation, such as the classical Maxwell/Möhr/Müller-Breslau virtual-work equations, extended to include terms for the inelastic behaviour. The solution of the equations was simplified if linear moment-rotation characteristics of the members of the frame were assumed. If the member between two plastic hinges was assumed to be rigid, the assumptions were identical with those of the Cambridge plastic-hinge method for a steel frame. This simplification could not yet be justified for reinforced concrete, since the ultimate inelastic deformations at hinges were often too small to allow full redistribution of bending moments to take place except in the

case of slabs. After more research, it might be possible to define certain framed structures to which the Cambridge method could be safely applied. The virtual-work equations, however, defined the essential compatibility not only at ultimate load but at any stage of loading. It was easy to adjust the reinforcement at critical sections of a concrete frame in order to vary the yield moment and ultimate resistance without altering the concrete section, a procedure which made it possible to design a frame satisfying the compatibility equations.

455. The method described in the Report for dealing with frame instability was not elegant but was safe and not too difficult to apply. Appropriate location of the plastic hinges reduced the problem to one of instability of a member, between hinges, subjected to sway and having a known end-restraint in bending.

456. Certain load-factors were recommended in the Report. It was assumed that factors less than 2 would be made only by the engineer responsible for the security of the structure. For competitive designs, the specification should be so drafted that it was not possible for an irresponsible designer-contractor to adopt a factor below 2 without the approval of a responsible engineer; there should be no competition in assessing load-factors to obtain an advantage in tendering.

457. The problem of shear was dealt with cautiously, since the mechanism of failure was not yet fully understood, and further research was required.

458. The basic assumptions governing the resistance of sections were considered only briefly since many investigations on this subject had been reported recently.

459. A slide was shown to illustrate the design rule, which was published in the Proceedings for November 1962, as an example of the type of simple rule which can be developed for use in certain cases and within well defined limits. The illustration which could be applied to a wide range of beams, was of a continuous beam, the live load on which did not exceed 50% of the dead load; there were also limiting ratios of span and values of parameters affecting crack-width and hinge-rotation. Comparison was made between the elastic solution and the ultimate-load solution, and it was seen that by employing the latter considerable economy resulted mainly because of the reduced amount of reinforcement required to resist the reduced bending moments. It was, however, necessary to check that cracks of excessive width did not occur at the potential plastic hinges under working load.

460. A second illustration showed the bending-moment distribution on the beam of a portal frame. Bending moments at the supports had been calculated for various stages of loading, temperature, creep and the like. For instance, the bending-moment distribution with the distributed load only at the beam was shown followed by the alteration caused by a small rise of temperature on the top of the beam. The next stage showed the bending-moment distribution after creep and shrinkage had taken place for a period of two days; it was remarked that this stage happened to coincide with the strict elastic solution. The subsequent diagrams gave the bending-moment distribution for a small rise of temperature at the bottom of the beam; there was no alteration for this condition after creep had taken place for two days, since the creep during the first two days was non-recoverable. The example showed clearly the variations possible under working-load conditions, and illustrated that the elastic solution was very approximate since considerable variations from the results of that solution could occur.

Summary of discussion

461. Some twelve speakers contributed to the discussion. Most dealt with more than one aspect of the Report, which was described by some as a sound statement of the results of research but, where it drifted into a code of practice, it was open to criticism. The principal matters discussed are collated and summarized in the following:

Load and load-factors

462. Some speakers expressed doubts regarding the accuracy of the magnitude of the applied loads especially when compared with the more exact knowledge of the load which would cause a structure to fail. Criticism was made of the loads specified in CP.3, Chapter 5. While speakers agreed that the assessment of the loading was outside the terms of reference of the Committee, the Institution was urged to investigate this matter so that loads could be defined more precisely and realistically.

463. Opinion of the load-factors recommended in the Report varied from complete disagreement to acceptance with some qualification. One speaker maintained that these load-factors were guesses and when combined with the guesses of engineers using the factors, there was no rational basis for selecting a suitable factor; the safeguards when using a factor of less than two were not adequately specified. Another speaker questioned the safety of low load-factors in view of the fact that, although limit design might be safe for a single application of a load, failure might occur at a lower load acting repeatedly.

464. The question of what was meant by 'failure' was raised, since this term plays so essential a part in the method advocated. Was it cessation of serviceability? Or collapse? The statement that failure moments could be calculated to within 2 or 3% was countered by the fact that in practice the strength of concrete cubes could vary considerably, and histograms relating to the formula for calculating the failure load on a beam had been displayed which showed results varying by several hundred per cent. Data giving the ratio of the calculated collapse load to the actual collapse load of a number of structures would be of value.

Stability

465. A number of speakers dealt at length with the subject of stability of frames. One criticized the method in the Report since all deflexions would become infinite for which condition the critical load on a column was 0·03 of the Euler load; to avoid this condition, the working load should not exceed 0·006 of the Euler load, compared with a typical value of 0·15 obtained by other methods of design of reinforced concrete columns. Slides were shown of a frame devised by the Building Research Station with joints restrained or released in various ways to show the considerable variation in the buckling load of one of the stanchions and to demonstrate particularly the extremely small load required to cause buckling when, in effect, numerous hinges had developed throughout the frame when free to sway. Another speaker considered this demonstration invalid, contending that it was not with wire frames that they were dealing but with frames having members of substantial stiffness, and with hinges that were not true hinges but having a definite moment of resistance.

466. The statement made by the Committee that, if one hinge less than the number sufficient for collapse is provided, stability is assured, was criticized on the grounds that a frame would not be stable in this condition. The proposal in the Report established a condition of equilibrium but, it was alleged it was not a condition of stable equilibrium. It was questioned whether the method in the Report could be applied and whether it had been applied to a large number of structures. The basis of the method was to limit deflexions and in this respect it might be a practicable method.

Redistribution of bending moments

467. Beyond the elastic range, the distribution of moments conformed to different laws, and only by knowing these laws could the true factor of safety be assessed. The speaker making this remark added that it was wrong to assume that the distribution of moments at the plastic stage was the same as at the elastic stage, and it could be shown easily that plasticity might act unfavourably since failure might occur with a moment diagram quite different from that assumed. Another aspect not well known was

that the small elastic deformations due to acceptable cracks could cause large re-distribution of moments at working load; the re-distribution might conform to that achieved at failure. Thus at working load, the distribution of the moments might be considerably different from that predicted by the elastic analysis. As another speaker pointed out, the efficiency of the detailing of the reinforcement had great effect on the distribution of cracks and therefore on the moments to be resisted. It is possible to reinforce a concrete structure, a later speaker said, in such a way that it could act in the way the designer required; present day research seemed to concentrate too much on finding out what the structure would do without taking this control into account.

468. Whether or not full redistribution of moments was achieved was discussed, one of several speakers quoting the results of tests in which full re-distribution was achieved in all instances. Another suggested that tests on concrete structures, especially the common two-span continuous beam, seldom covered cases of severe rotation. It was demonstrated that the type of load and the type of structure had important effects on the rotation required, and concrete could not always sustain the large rotation necessary in some cases to achieve full equalization of moments. Eventually a simple system for checking ultimate-load designs might be evolved in which permissible ratios of ultimate moments at midspan and support could be stated for different loadings and conditions of continuity taking into account the strength of the concrete and the proportions of tension and compression reinforcement.

Rotations

469. Whether an actual structure would exhibit the same distribution of moments as considered in the design analysis depends on the accuracy of the assumed moment-curvature diagrams. This speaker pointed out that, to ensure that the design method was safe, the Report limited the rotation to a certain value such that at every section of the structure the moment would lie on the ascending portion of the moment-curvature curve; moments exceeding this value might result in sudden failure. If hinge rotations were limited and ultimate moments were underestimated, safe but conservative design followed. However, neither rotation nor moments at critical sections of a frame designed by such a method had any relation to those actually occurring in a highly redundant frame at collapse.

470. The amount of reinforcement had a determining effect on the rotational characteristics. One speaker showed moment-rotation diagrams for various members demonstrating the smaller ultimate rotation which obtains when the proportion of reinforcement is high. Another speaker showed that the smaller the depth to the neutral axis, the more a plastic hinge could rotate before failure; this condition could be achieved in an under-reinforced beam, but this might not always be convenient since the bending strength might be too small. The provision of compression reinforcement, however, caused a reduction in the neutral-axis depth and consequent increase of rotational capacity; this method was economic. It was stated by another speaker that, if the compression zone of a beam was restrained by helical bending (see *Concrete and Constructional Engineering*, December 1962), the ultimate moment was practically constant for a wide range of curvatures. This speaker also explained that considerable research had been done with regard to the moment-curvature relation of beams but with very little regard to frames; further research might result in simplification of design procedure. Other factors such as the restraining action of floors and walls had to be considered and could modify considerably the idealized pattern of hinges assumed in a skeletal frame.

Comparison of concrete and steel structures

471. The statement regarding the alleged inapplicability of the Cambridge method (relating to the ultimate-load design of steel structures) to concrete structures led to comments from several speakers. One pointed out that the ultimate load on a reinforced concrete structure was determined by the deformation whereas on a steel

structure it was determined mainly by equilibrium. Another speaker explained that the Cambridge method could be used so long as the deflexions could be checked, and elaborated an analytical sequence, but one of the disadvantages of the analysis was that the collapse solution implied $(N+1)$ hinges having been formed, whereas it was necessary to consider (as in the Report) the condition of the concrete frame when N hinges only had formed. Also the rotations would be too large and adjustment of the bending moments impracticable, necessitating a complete re-working of the problem.

472. The method given in the Report, stated one speaker, was an adaptation of the method developed for steel structures using the same idealized elastic-plastic moment-curvature relationship but limiting rotations to permissible values; it did not give an accurate prediction of the behaviour of a frame at ultimate load because it ignored the fact that the stress-strain curve of concrete rose to a maximum and then descended. It was shown that there was no sudden failure of concrete and large strains could occur without becoming structurally useless. Consequently curvatures much larger than that occurring at the maximum moment were possible. The strain of 0·0035 recommended in the Report was safe for dry mature concrete. Since concrete was a strain-softening material whereas structural steel was a strain-hardening material, concepts used in the plastic design of steel structures could not be applied to concrete structures. More research was required on the behaviour of concrete under large curvature with maximum moments before a rational limit and design could be produced and, as another speaker maintained, further research was necessary especially in view of the fact that there was as yet no positive criteria for the failure of concrete. Another speaker had asked what tests had been carried out to prove the Cambridge method was not applicable to concrete, and the reply had been that concrete was too 'brittle'; this statement was disproved, a further speaker alleged, by the foregoing considerations although the strain and collapse characteristics of concrete depend, as yet another explained, on whether the load was applied rapidly or on a long-term basis.

Slabs

473. It was thought by some speakers that the Report did not give to slabs the attention they merited, and even in the little consideration given there was some confusion between special and general cases. It was inferred that serviceability of slabs need not be considered, but one speaker suggested that, since the ultimate-load solution was generally in terms of the strength in two directions and since the designer had to choose the ratio of these strengths, a "serviceable" slab is produced by selecting a ratio equal to that of the ratio between the elastic moments.

474. One speaker, in dealing with the acceptance of the yield-line theory in the Report, suggested that verification of this theory only applied to simple cases. In a specific case, the actual form at destruction of the slab was very different from that predicted by the theory. Another speaker brought to notice the fact that the yield-line theory had been used for the design of thousands of sound slabs throughout the world, and this might be an acceptable substitute for tests. The fact that the Report did not deal fully with the theory meant that designers wishing to take the matter further should consult appropriate books. The collapse method of design is the only practical method except for slabs of simple shapes.

Shear

475. One speaker stated that the Report gave little guidance on how to deal with bending combined with shearing. One point raised was that, if designed by the ultimate-load method, was the bending strength only taken into account and the shear disregarded? If so it was folly. If considering shear resistance elastically were bending effects ignored? The settlement of these matters was important in a member subjected to high shearing forces.

General

476. Many contributors expressed succinct remarks regarding ultimate-load design, among which were the following.

477. It was erroneous to consider the ultimate-load method as a substitute for the elastic theory; the former predicted the failure of the structure but had to be applied simultaneously with the latter.

478. Calculation at the plastic stage was justified in all cases because plasticity was not an additional safety factor compared with the elastic calculation, but was a particular stage in the behaviour of a structure. The properties of a structure in that zone had to be known and taken into account if safety margins fixed previously were to be obtained.

479. From the point of view of rationalizing structural design, the ultimate-load method had a great advantage over elastic theories since the ultimate moment could be predicted much more accurately than the yield moment. The Report assumed that the predicted ultimate moment was less than the actual moment but did not give the amount by which it was less.

PROFESSOR BAKER'S REPLY

Professor Baker in replying briefly to the discussion said that although it was necessary to know much more about actual loading, this ignorance should not be a criticism of the Report for suggesting that the maximum load should be used and that it was the designer's responsibility to find out what the maximum load was. Regarding load-factors, there seemed no reason why a bridge, for instance, should have the same load-factor as a fence post. If the numerical grading of the load-factors given in the Report were considered, it would be seen that it was rational since all the factors one had in mind when thinking of the security of a structure were taken into account in as simple a quantitative manner as practicable. Experience showed that a load-factor of 2 was sufficient; the range was only from 1·25 to 2. If designers selected the load-factor for a given design from the table in the Report, they were likely to agree fairly well with the resulting design. It would be difficult to select a load-factor much below 2; that was intentional in order to safeguard against irresponsibility. The difference between the two schools of thought regarding the value of the load-factor compared with the load used in the calculations is that in one case the emphasis is on determining the maximum load. In the other a higher load-factor is used and a load closer to the average is assumed. It seems, however, to be much safer and more logical to determine the maximum load and to adopt a lower load-factor, rather than the reverse.

481. One speaker had rightly shown that sometimes the wrong test had been used to indicate that it was unnecessary to check rotations. Tests must be significant and should be those in which the worst cases of rotation occurred.

482. The Euler loads given by another speaker might apply to steel but might not apply to concrete. If one had a buckling or a swaying member of concrete and there was uncertainty about the curvature, the inelegant method suggested in the Report gave a means of analysis by first isolating storey from storey by placing hinges where one wanted them; one then checked to see whether there was a slender column in any storey or whether a beam was very slender. A maximum eccentricity due to all causes was then assumed and the deflexions were then calculated in the hope that they were less than the assumed deflexions. It was hoped that in due course a more elegant method of design might be produced for this complicated problem. The European Concrete Committee's empirical formula for the buckling of reinforced concrete columns, which was based on a large number of tests, should be of considerable help to designers, because once the position of the hinges had been selected, it was possible to deal with the column between the hinges. If the column tended to buckle, the Committee's formula could be applied, at least for the first trial calculations, to determine the size of the column; one then made a test of the assumed sway

and eccentricity due to buckling to prove that the calculated deflexion was less than the assumed deflexion.

483. The fact pointed out by another speaker that, even at working load, there could be considerable deviation from the elastic distribution of the bending moments due to inelastic effects is important in bridge design and has to be taken into account in order to reduce the cracking under working load.

484. Referring to tests which had shown complete redistribution of moments, these were not necessarily the worst cases, as was clear from the design rule incorporated in the Proceedings for November 1962. In that rule it is clearly seen that if the live load exceeds the dead load by a certain percentage, or if the spans varied by more than a certain amount, excessive rotations of the hinges occur. A rule of that kind could be used perhaps to decide which was the worst case likely to occur in practice, and to carry out a test accordingly.

485. The uniqueness of the hinge system, when checked by the compatibility equation, had been questioned (not in the discussion), and the matter that different cases of live load might give different positions of hinges had also been raised; it was for the designer to judge whether it was necessary to investigate these alternative positions. As to the uniqueness of the system, it was not so easy to deal with it briefly without having had time to consider the problem. In the idealized case, a linear relationship of moment to rotation or to curvature up to a certain limit was assumed, an assumption fairly close to reality. At the other extreme, there might be an imaginary case in which hinges formed at positions different from the final positions and formed at an early stage of loading and rotated in the opposite direction to the final direction of rotation. In such a case it would be necessary for rotational characteristics to be in the negative direction of the moment. Assuming linear moment-rotation characteristics applying to this odd case and having a load about equal to the working load, the condition of negative rotation and negative moment is unlikely to result in the member entering the plastic zone. If it did, and when the positive moment developed, the only effect would be that one had displaced the moment-rotation curve slightly to the left. It could be shown that the error resulting from such a permanent set must be small. Reference should be made to an article in *Concrete and Constructional Engineering* for March 1961; if the methods described in that article were applied to the worst cases of a beam, it would be found that the error could not be very large even though such an extraordinary action happened. In practical designs which would be checked for serviceability, it would be almost impossible for a hinge to rotate in the wrong direction and even if it did happen, the result would not be serious in regard to the ultimate resistance.

486. With regard to the drooping characteristics of a stress-strain curve, it was possible also to obtain a rising characteristics depending on the rate of loading; reference should be made to the tests by Professor Rüsch, which were reported in a recent issue of the *Journal of the American Concrete Institute* and which constituted a comprehensive study of the influence of the rate of loading in compression tests. In the method of design advocated in the Research Committee's Report, drooping characteristics were not considered, since the limiting unbound concrete strain was 0·0035.

487. In connexion with the influence of binding there was a brief reference in the Report to the increased strains which could be allowed if binding was provided.

488. As emphasized in the introductory remarks, it was important to consider a structure at all stages of loading. Designers could not be divided into two schools: the elastic school and the inelastic school.

ISBN
0 901948 64 0